Non-Timber Forest Products: Medicinal Herbs, Fungi, Edible Fruits and Nuts, and Other Natural Products from the Forest

Non-Timber Forest Products: Medicinal Herbs, Fungi, Edible Fruits and Nuts, and Other Natural Products from the Forest has been co-published simultaneously as *Journal of Sustainable Forestry*, Volume 13, Numbers 3/4 2001.

The *Journal of Sustainable Forestry* Monographic "Separates"

Below is a list of "separates," which in serials librarianship means a special issue simultaneously published as a special journal issue or double-issue *and* as a "separate" hardbound monograph. (This is a format which we also call a "DocuSerial.")

"Separates" are published because specialized libraries or professionals may wish to purchase a specific thematic issue by itself in a format which can be separately cataloged and shelved, as opposed to purchasing the journal on an on-going basis. Faculty members may also more easily consider a "separate" for classroom adoption.

"Separates" are carefully classified separately with the major book jobbers so that the journal tie-in can be noted on new book order slips to avoid duplicate purchasing.

You may wish to visit Haworth's website at . . .

http://www.HaworthPress.com

. . . to search our online catalog for complete tables of contents of these separates and related publications.

You may also call 1-800-HAWORTH (outside US/Canada: 607-722-5857), or Fax 1-800-895-0582 (outside US/Canada: 607-771-0012), or e-mail at:

getinfo@haworthpressinc.com

Non-Timber Forest Products: Medicinal Herbs, Fungi, Edible Fruits and Nuts, and Other Natural Products from the Forest, edited by Marla R. Emery, Rebecca J. McLain (Vol. 13, No. 3/4, 2001). *Focuses on NTFP use, research, and policy concerns in the United States. Discusses historical and contemporary NTFP use, ongoing research on NTFPs, and socio-political considerations for NTFP management.*

Understanding Community-Based Forest Ecosystem Management, edited by Gerald J. Gray, Maia J. Enzer, and Jonathan Kusel (Vol. 12, No. 3/4 & Vol. 13, No. 1/2, 2001). *Here is a state-of-the-art reference and information source for scientists, community groups and their leaders, resource managers, and ecosystem management practitioners. Healthy ecosystems and community well-being go hand in hand, and the interdependence between the two is the focal point of community-based ecosystem management. The information you'll find in* ***Understanding Community-Based Forest Ecosystem Management*** *will be invaluable in your effort to manage and maintain the ecosystems in your community.* ***Understanding Community-Based Forest Ecosystem Management*** *examines the emergence of community-based ecosystem management (CBEM) in the United States. This comprehensive book blends diverse perspectives, enabling you to draw on the experience and expertise of forest-based practitioners, researchers, and leaders in community-based efforts in the ecosystem management situations that you deal with in your community.*

Climate Change and Forest Management in the Western Hemisphere, edited by Mohammed H. I. Dore (Vol. 12, No. 1/2, 2001). *This valuable book examines integrated forest management in the Americas, covering important global issues including global climate change and the conservation of biodiversity. Here you will find case studies from representative forests in North, Central, and South America. The book also explores the role of the Brazilian rainforest in the global carbon cycle and implications for sustainable use of rainforests, as well as the carbon cycle and the valuation of forests for carbon sequestration.*

Mapping Wildfire Hazards and Risks, edited by R. Neil Sampson, R. Dwight Atkinson, and Joe W. Lewis (Vol. 11, No. 1/2, 2000). *Based on the October 1996 workshop at Pingree Park in Colorado,* ***Mapping Wildfire Hazards and Risks*** *is a compilation of the ideas of federal and state agencies, universities, and non-governmental organizations on how to rank and prioritize forested watershed areas that are in need of prescribed fire. This book explains the vital importance of fire for the health and sustainability of a watershed forest and how the past acceptance of fire suspension has consequently led to increased fuel loadings in these landscapes that may lead to more severe future wildfires. Complete with geographic maps, charts, diagrams, and a list of locations where there is the greatest risk of future wildfires,* ***Mapping Wildfire Hazards and Risks*** *will assist you in deciding how to set priorities for land treatment that might reduce the risk of land damage.*

Frontiers of Forest Biology: Proceedings of the 1998 Joint Meeting of the North American Forest Biology Workshop and the Western Forest Genetics Association, edited by Alan K. Mitchell, Pasi Puttonen, Michael Stoehr, and Barbara J. Hawkins (Vol. 10, No. 1/2 & 3/4, 2000). *Based on the 1998 Joint Meeting of the North American Forest Biology Workshop and the Western Forest Genetics Association, Frontiers of Forest Biology addresses changing priorities in forest resource management. You will explore how the emphasis of forest research has shifted from productivity-based goals to goals related to sustainable development of forest resources. This important book contains fascinating research studies, complete with tables and diagrams, on topics such as biodiversity research and the productivity of commercial species that seek criteria and indicators of ecological integrity.*

"There is clear emphasis on the genetics, genecology, and physiology of trees, particularly temperate trees. . . . These proceedings are also testimony to what does or should distinguish forest biology from other sciences: a focus on intra- and inter-specific interactions between forest organisms and their environment, over scales of both time and place." (Robert D. Guy, PhD, Associate Professor, Department of Forest Sciences, University of British Columbia, Vancouver, Canada)

Contested Issues of Ecosystem Management, edited by Piermaria Corona and Boris Zeide (Vol. 9, No. 1/2, 1999). *Provides park rangers, forestry students and personnel with a unique discussion of the premise, goals, and concepts of ecosystem management. You will discover the need for you to maintain and enhance the quality of the environment on a global scale while meeting the current and future needs of an increasing human population. This unique book includes ways to tackle the fundamental causes of environmental degradation so you will be able to respond to the problem and not merely the symptoms.*

Protecting Watershed Areas: Case of the Panama Canal, edited by Mark S. Ashton, Jennifer L. O'Hara, and Robert D. Hauff (Vol. 8, No. 3/4, 1999). *"This book makes a valuable contribution to the literature on conservation and development in the neo-tropics. . . . These writings provide a fresh yet realistic account of the Panama landscape." (Raymond P. Guries, Professor of Forestry, Department of Forestry, University of Wisconsin at Madison, Wisconsin)*

Sustainable Forests: Global Challenges and Local Solutions, edited by O. Thomas Bouman and David G. Brand (Vol. 4, No. 3/4 & Vol. 5, No. 1/2, 1997). *"Presents visions and hopes and the challenges and frustrations in utilization of our forests to meet the economical and social needs of communities, without irreversibly damaging the renewal capacities of the world's forests." (Dvoralai Wulfsohn, PhD, PEng, Associate Professor, Department of Agricultural and Bioresource Engineering, University of Saskatchewan)*

Assessing Forest Ecosystem Health in the Inland West, edited by R. Neil Sampson and David L. Adams (Vol. 2, No. 1/2/3/4, 1994). *"A compendium of research findings on a variety of forest issues. Useful for both scientists and policymakers since it represents the combined knowledge of both." (Abstracts of Public Administration, Development, and Environment)*

∞ ALL FOOD PRODUCTS PRESS BOOKS
AND JOURNALS ARE PRINTED
ON CERTIFIED ACID-FREE PAPER

Non-Timber Forest Products: Medicinal Herbs, Fungi, Edible Fruits and Nuts, and Other Natural Products from the Forest

Marla R. Emery
Rebecca J. McLain
Editors

Non-Timber Forest Products: Medicinal Herbs, Fungi, Edible Fruits and Nuts, and Other Natural Products from the Forest has been co-published simultaneously as *Journal of Sustainable Forestry*, Volume 13, Numbers 3/4 2001.

Food Products Press
An Imprint of
The Haworth Press, Inc.
New York • London • Oxford

Published by

Food Products Press®, 10 Alice Street, Binghamton, NY 13904-1580 USA

Food Products Press® is an imprint of The Haworth Press, Inc., 10 Alice Street, Binghamton, NY 13904-1580 USA.

Non-Timber Forest Products: Medicinal Herbs, Fungi, Edible Fruits and Nuts, and Other Natural Products from the Forest has been co-published simultaneously as *Journal of Sustainable Forestry*, Volume 13, Numbers 3/4 2001.

The development, preparation, and publication of this work has been undertaken with great care. However, the publisher, employees, editors, and agents of The Haworth Press and all imprints of The Haworth Press, Inc., including The Haworth Medical Press® and Pharmaceutical Products Press®, are not responsible for any errors contained herein or for consequences that may ensue from use of materials or information contained in this work. Opinions expressed by the author(s) are not necessarily those of The Haworth Press, Inc.

Cover design by Thomas J. Mayshock Jr.

Library of Congress Cataloging-in-Publication Data

Non-timber forest products: medicinal herbs, fungi, edible fruits and nuts, and other natural products from the forest/Marla R. Emery, Rebecca J. McLain, editors.

p. cm.

Co-published simultaneously as Journal of Sustainable Forestry, volume 13, Numbers 3/4 2001.

Includes bibliographical references (p.).

ISBN 1-56022-088-0 (alk. paper)–ISBN 1-56022-089-9 (pbk.: alk. paper)

1. Non-timber forest products–United States. 2. Non-timber forest products. I. Emery, Marla R. II. McLain, Rebecca J. (Rebecca Jean)

SD543.3.U6 N66 2001
634.9′87′0973–dc21 2001040229

Indexing, Abstracting & Website/Internet Coverage

This section provides you with a list of major indexing & abstracting services. That is to say, each service began covering this periodical during the year noted in the right column. Most Websites which are listed below have indicated that they will either post, disseminate, compile, archive, cite or alert their own Website users with research-based content from this work. (This list is as current as the copyright date of this publication.)

Abstracting, Website/Indexing Coverage	Year When Coverage Began
• *Abstract Bulletin*	1993
• *Abstracts in Anthropology*	1993
• *Abstracts on Rural Development in the Tropics (RURAL)*	1993
• *AGRICOLA Database*	1993
• *Biology Digest (in print & online)*	2000
• *Biostatistica*	1993
• *BUBL Information Service, an Internet-based Information Service for the UK higher education community <URL: http://bubl.ac.uk/>*	1995
• *CNPIEC Reference Guide: Chinese National Directory of Foreign Periodicals*	1996
• *Engineering Information (PAGE ONE)*	1999
• *Environment Abstracts <www.cispubs.com>*	1993
• *Environmental Periodicals Bibliography (EPB)*	1993
• *FINDEX <www.publist.com>*	1999

(continued)

- *Forestry Abstracts (c/o CAB Intl/CAB ACCESS) <www.cabi.org>* **1993**
- *GEO Abstracts (GEO Abstracts/GEOBASE) <URL: http://www.elsevier.nl>* **2000**
- *Human Resources Abstracts (HRA)* **1993**
- *Journal of Planning Literature/Incorporating the CPL Bibliographics* **1995**
- *Referativnyi Zhurnal (Abstracts Journal of the All-Russian Institute of Scientific and Technical Information-in Russian)* .. **1993**
- *Sage Public Administration Abstracts (SPAA)* **1993**
- *Sage Urban Studies Abstracts (SUSA)* **1993**
- *Wildlife Review Abstracts <www.nisc.com>* **1993**

Special Bibliographic Notes related to special journal issues (separates) and indexing/abstracting:

- indexing/abstracting services in this list will also cover material in any "separate" that is co-published simultaneously with Haworth's special thematic journal issue or DocuSerial. Indexing/abstracting usually covers material at the article/chapter level.
- monographic co-editions are intended for either non-subscribers or libraries which intend to purchase a second copy for their circulating collections.
- monographic co-editions are reported to all jobbers/wholesalers/approval plans. The source journal is listed as the "series" to assist the prevention of duplicate purchasing in the same manner utilized for books-in-series.
- to facilitate user/access services all indexing/abstracting services are encouraged to utilize the co-indexing entry note indicated at the bottom of the first page of each article/chapter/contribution.
- this is intended to assist a library user of any reference tool (whether print, electronic, online, or CD-ROM) to locate the monographic version if the library has purchased this version but not a subscription to the source journal.
- individual articles/chapters in any Haworth publication are also available through the Haworth Document Delivery Service (HDDS).

Non-Timber Forest Products: Medicinal Herbs, Fungi, Edible Fruits and Nuts, and Other Natural Products from the Forest

CONTENTS

ABOUT THE EDITORS

Marla R. Emery, PhD, is Research Geographer with the Northeastern Research Station of the USDA Forest Service, where her research focuses on the role of non-timber forest products (NTFPs) in household economies and other direct human-forest interactions. She conducted the first comprehensive study of contemporary NTFP use in the United States, for which she spent a year in Michigan's Upper Peninsula conducting ethnographic research that documented the material uses of 138 products from over 80 botanical species and the livelihood practices associated with them. She is currently repeating that work in the northeastern United States as well as conducting research on fine-scale land use in the Adirondack Park region of New York. Her past duties with the Forest Service have included developing an agenda for research on the human dimensions of global environmental change for the Forest Service's Northern Global Change Program. Dr. Emery came to the Forest Service from the National Research Council (NRC) in Washington, DC, where she served as Staff Officer for the U.S. National Committee for the International Decade for Natural Disaster Reduction. During her four years at the NRC, she worked extensively with international organizations and agencies of the U.S. Federal Government. She also spoke to groups in the United States and abroad about natural disaster reduction. Before joining the staff of the NRC she worked for eight years as an educator. Dr. Emery has a BA in French/Spanish from San José State University, California, a Master's of Science in Education from the University of Miami, Florida, and a PhD in Geography from Rutgers University in New Jersey.

Rebecca J. McLain, PhD, has an interdisciplinary background that supports her life-long interest in exploring human-environmental interactions. Her formal education includes a BA in Cultural Anthropology (U. Texas-Austin), an MS in Land Resources (U. Wisconsin-Madison), and a PhD in Forest Policy (U. Washington-Seattle). Her field experience includes managing applied forest policy research programs for the U. Wisconsin Land Tenure Center in West Africa and Haiti, and several program management positions with natural resource management agencies in the United States, including cultural resources management with the Bureau of Land Management, environmental education with the Washington State Department of Ecology, and agricultural impact analysis with the State of Wisconsin. She has recently completed a long-term ethnographic study of wild mushroom politics in central Oregon, and is coordinating the publication of an edited volume on non-timber forest products management issues and concerns in the United States. She is co-director of the Institute for Culture and Ecology, a 501(c)3 organization based in Portland, Oregon that seeks to bridge the gap between academic researchers and natural resource management practitioners.

Introduction

Rebecca J. McLain
Marla R. Emery

Wherever and whenever forests and humans have occupied the same space on Earth, it can be expected that Non-Timber Forest Products (NTFP)[1] have made important contributions to people's livelihoods. NTFP research and policy, however, have generally focused on the Third World. This special issue shifts attention to NTFP use, research, and policy concerns in the United States as a way of illustrating the important contribution of these products to post-industrial societies.

Although NTFPs are often overlooked by public and private forest land managers, the contributions in Section I illustrate that NTFP use and management in the Pacific Northwest and Upper Midwest has a very long history and continues to be widespread, complex and dynamic. Thadani begins the discussion and provides important comparative background in a paper on the development of NTFPs as management and conservation strategies in the Third World. Questions raised by his work include: What are the differences and similarities between NTFPs' importance to rural residents of the tropics and the United States? What social, economic, and ecological difficulties lurk in the promotion of NTFP commercialization as a sustainable development

Rebecca J. McLain is Co-Founder and Director of the Institute for Culture and Ecology, P.O. Box 6688, Portland, OR 97228 (E-mail: mclain@ifcae.org).

Marla R. Emery is Research Geographer with the USDA Forest Service, Northeastern Research Station, Burlington, VT 05402-0968 (E-mail: memery@fs.fed.us).

[Haworth co-indexing entry note]: "Introduction." McLain, Rebecca J., and Marla R. Emery. Co-published simultaneously in *Journal of Sustainable Forestry* (Food Products Press, an imprint of The Haworth Press, Inc.) Vol. 13, No. 3/4, 2001, pp. 1-4; and: *Non-Timber Forest Products: Medicinal Herbs, Fungi, Edible Fruits and Nuts, and Other Natural Products from the Forest* (ed: Marla R. Emery, and Rebecca J. McLain) Food Products Press, an imprint of The Haworth Press, Inc., 2001, pp. 1-4. Single or multiple copies of this article are available for a fee from The Haworth Document Delivery Service [1-800-342-9678, 9:00 a.m. - 5:00 p.m. (EST). E-mail address: getinfo@haworthpressinc.com].

strategy? and How might lessons from the developing world inform research and management in post-industrial settings? Emery and O'Halek then set the stage for a discussion of present-day NTFP issues in the United States by describing the historical context of NTFP use and management in the Pacific Northwest and Upper Midwest. Among the questions they address are: What role have NTFPs played in different cultures within the United States, what products have been important regionally and nationally, and how has the economic importance of NTFPs varied by region and time period over the past several hundred years? Next, Turner and Cocksedge describe aboriginal (e.g., First Nations and Native American) uses of NTFPs on both sides of the western Canada-U.S. border. Their monograph suggests several questions: What are the distinctive interests of aboriginal peoples in NTFP use and management throughout North America? What is the relationship between subsistence and trade uses of NTFPs? What ecological and social factors influence the sustainability of harvesting? Alexander and McLain follow this analysis with a brief examination of economic trends in three major NTFP sub-sectors–medicinals, floral greens, and wild edibles. Their discussion addresses the questions: What species and products are harvested in the United States today? What is the economic scope of specific NTFP markets? How are NTFP markets and sources of supply in the United States tied into global economies? What are the consequences of increased demand for NTFP products and species for the sustainability of forested ecosystems? Freed concludes Section I with his case study of a county in the state of Washington, illustrating the role of NTFPs in local economies.

The production of a more comprehensive base of scientific knowledge for NTFPs is often identified as a critical component for sustainable management of NTFPs under conditions of industrial extraction. NTFP research in the United States is poorly funded, fragmented and limited in scope in comparison to research on timber, recreation, and wildlife. However, networks of scientific researchers interested in NTFPs are beginning to form in the United States and Canada. The densest node of scientific activity on NTFPs exists in the Pacific Northwest region, an area with a large supply of a variety of commercially valuable NTFPs and a highly contentious forest management context. In Section II, Vance, Pilz et al., and Alexander et al. provide an overview of this scientific activity in their discussions of different aspects of the USDA Forest Service Pacific Northwest Research Station's NTFP research programs and projects. Their contributions focus on the following questions: What is the scope and content of on-going and proposed NTFP research programs? How do these programs address on-the-ground forest management concerns? Which scientific disciplines are incorporated into NTFP research? What attempts are scientists making to integrate NTFP research across disciplines and to link a variety of forest stakeholders into the development and implementation of NTFP research agendas?

The participation of NTFP resource users in policy making and implementation also is often cited as a necessary component of sustainable NTFP management. In Section III, Love and Jones, Emery, Hansis et al., and McLain and Jones address questions related to the existing and potential roles that NTFP harvesters and buyers play in managing NTFPs sustainably. Some of these questions include: Why have NTFPs become a policy and management issue at this moment in history? What types of knowledge do harvesters have of NTFPs and the environments in which they are located? What stewardship practices do harvesters engage in? How is the social composition of harvester populations changing as demand for products increases in certain parts of the United States? What are some of the key characteristics of NTFP tenure regimes, and what conflicts have arisen as the socio-ecological context within which these were developed has changed? To what extent are harvesters and buyers involved in forest management decisions? What factors limit their involvement? And what efforts are they making to expand their political influence?

McLain and Alexander end the issue with a summary of the lessons learned from the research described earlier, noting in particular the importance of encouraging collaborative and interdisciplinary types of research. They also point out the relevance of NTFP research in the United States to the work being done on similar issues in other parts of the world, and thus the importance of widening and strengthening the global scope of NTFP research networks.

This issue represents the joint efforts of more than a dozen researchers with training and experience in disciplines as diverse as anthropology, ecology, economics, forestry, geography, mycology, and policy science. We share a belief that NTFPs play important roles in ecological, economic, and cultural systems. We also have in common a dedication to the expansion and development of informal and formal research networks that facilitate the exchange of information about

NTFPs and encourage innovative thinking about the current and potential roles of NTFPs in socio-ecological systems.

With the exception of one contributor, whose research focuses on NTFP harvesters in Northern Michigan, the contributors to this issue work and live in the Pacific Northwest region of northern North America. As a result, many of the articles and case examples focus on research or issues relevant to that region. The dynamics of NTFP use and management may be quite different in other parts of the United States and Canada. We hope that this issue will encourage the development of comparative work on NTFPs within the United States and Canada, as well as between northern North America and other parts of the world. We believe this issue constitutes a useful addition to the fine body of existing work on NTFPs and hope that it will encourage both researchers and policy makers to think of NTFPs as a truly global issue.

NOTE

1. The terms "non-timber forest products" (NTFP), "non-wood forest products" (NWFP), and "special forest products" (SFP) are used interchangeably in this issue. Although the term "non-timber forest products" is gaining popularity, many people in the United States use the term "special forest products" when referring to forest products such as berries, fungi, leaves, boughs, roots, and bark.

SECTION I: NON-TIMBER FOREST PRODUCTS, PAST AND PRESENT

International Non-Timber Forest Product Issues

Rajesh Thadani

SUMMARY. Perceived as socially, economically, and ecologically sustainable, non-timber forest products (NTFPs) have held special charm as alternatives to forest management focussed exclusively on timber. This paper examines themes central to development of NTFPs as management and conservation strategies in the developing world. Following brief descriptions of seven product types, the paper reviews research on the promise of sustainable prosperity through NTFPs. Critiques of economic valuation and commercialization suggest that NTFP development strategies are not without social, economic, and ecological problems. The paper concludes with a list of eight major issues relating to the extraction

Rajesh Thadani is Coordinator of Central Himalayan Rural Action Group (Chirag), Sitla, P.O. Mukteswar, District Nainital, U.P. 263 138, India (E-mail: rajesh.thadani@yale.edu).

Critical comments provided by Dr. Mark S. Ashton are acknowledged.

[Haworth co-indexing entry note]: "International Non-Timber Forest Product Issues." Thadani, Rajesh. Co-published simultaneously in *Journal of Sustainable Forestry* (Food Products Press, an imprint of The Haworth Press, Inc.) Vol. 13, No. 3/4, 2001, pp. 5-23; and: *Non-Timber Forest Products: Medicinal Herbs, Fungi, Edible Fruits and Nuts, and Other Natural Products from the Forest* (ed: Marla R. Emery, and Rebecca J. McLain) Food Products Press, an imprint of The Haworth Press, Inc., 2001, pp. 5-23. Single or multiple copies of this article are available for a fee from The Haworth Document Delivery Service [1-800-342-9678, 9:00 a.m. - 5:00 p.m. (EST). E-mail address: getinfo@haworthpressinc.com].

and trade of non-timber forest products.

KEYWORDS. Extractive reserves, economic valuation, commercialization, third world, non-timber forest products

INTRODUCTION

Forest outputs can be divided into three categories–timber, non-timber and environmental services. While, the former two are consumptive uses of the forest, environmental services together with uses such as eco-tourism are non-consumptive. Traditionally, the value of a forest has been determined by its potential consumptive uses, in particular the value of timber. For over a century, the model by which forests have been managed around the world has focussed largely on timber maximization while other forest outputs, such as non-timber forest products (NTFPs) and environmental services, have been ignored or regarded as byproducts of the timber production process (Tewari, 1998, p. 1). Timber extraction, however, often has severe consequences for forest structure and its ability to provide both non-timber products and environmental services. Over the past two decades the value of forests as regulators of climate patterns and flood cycles, areas of pollution abatement, and centers of tourism have been increasingly stressed. In this regard, the collection of NTFPs, a consumptive use of the forest, but one that allows the preservation of forest structure and ecosystem integrity, has increasingly been studied as an alternate to logging.

Non-timber forest products include a myriad of products that have been extracted from forests around the world for millennia. These include fuel and fodder biomass, construction materials like canes and rattan, fiber, flowers, fruits, tubers, edible and hallucinogenic mushrooms, seeds, oils, medicinal plants, gums and resins, tannins and dyes, and a variety of birds, mammals, fish and insects used as food or sold to collectors as well as wildlife products such as honey and eggs. As these products are most commonly consumed locally, early economic evaluations of forests failed to register the importance of NTFPs or 'minor forest products' as they were formerly called. A theory said to have originated in Central Europe, the so-called 'wake theory' argued that these minor forest products would be produced automatically in the course of timber production, much as a forward moving ship produces a wake through the water (Freezailah, 1992). Such beliefs are no longer tenable. A benchmark paper by Peters et al. (1989) indicated that sustained yield extraction might be more profitable than other major land uses that destroy forests. While Peters et al. (1989) did not try to generalize their results beyond the forests they studied, many later researchers did, and the old paradigm that rain forests in their intact state were worth less than when they were destroyed began to be reversed (Allegretti, 1990).

While many of the studies that highlighted the potential of non-timber forest products originated in the Amazon basin, data on their economic worth was becoming available from other parts of the world. In India, for example, this recognition came in the 1970's when the Government found that revenues from NTFPs had increased tremendously over time and were comparable to the revenues accrued from timber sales. This led to the establishment of state level Forest Development Corporations (FDCs) to manage the collection and marketing of NTFPs (Tewari, 1998). According to one estimate, export earnings from NTFPs account for 60-70% of the total export earnings from forest products in India, and this proportion is rising (Tewari and Campbell, 1995). In Indonesia, exports of NTFPs were worth U.S. $120 million in 1982, an amount almost half that of government revenues from timber production (Panayotou and Ashton, 1992). By 1994, Indonesia's rattan exports alone were worth $360 million (de Beer and McDermott, 1996).

Impetus for finding economic value for NTFPs was provided by an unusually long dry season experienced in the Amazon basin in 1987, which resulted in the largest scale burn recorded in the history of that region. The annual deforestation rate for the region from 1981-88 was 1.2%, a four-fold increase over the rate from 1976-78 (Anderson, 1990a). This, coupled with an exceptionally severe drought in the United States in 1988, served to bring tropical deforestation in Amazonia into focus. Issues such as global warming and tropical deforestation, once perceived as remote problems of marginal significance were transformed into major environmental issues (Anderson, 1990a). The extraction of timber, was found usually to result in permanent forest destruction in the tropics. Lanly (1982) approximated that only 4% of the tropical forests were subject to some level of harvesting regulations designed to promote regeneration. Poore (1989, p. 196) found less than 0.12% of tropical forests to be under management for sustained yields of timber. The perception that NTFP extraction could be a sustainable use of the forests that would simultaneously benefit indigenous people and generate revenues comparable to timber extraction, greatly increased the charm associated with non-timber forest products. In 1988 the International Tropical Timber Organization (ITTO) published a report calling for the rigorous study of NTFPs (Panayotou and Ashton, 1992). In a single notable year, 1989, several significant evaluations of NTFPs were published (de Beer and McDermott, 1989; Godoy and Feaw, 1989, Peters et al. 1989).

NTFPs are extracted in forests in every country around the world. However, it is in the developing countries, the 'Third World,' where the study of NTFPs has received the most attention, as dependence on NTFPs may be highest among people of these countries. NTFPs are a critical component for the sustenance of nearly 500 million people living in and around the forests in India (World Resources Institute, 1990). Of these, an estimated 50 million tribals are largely dependant on NTFPs (Poffenberger, 1996). One-fifth of the economically active population of the forested zone of southern Ghana earn part of their income from NTFPs (Ndoye et al. 1998). Many of these countries lie in the tropics and sub-tropics, the luxuriant forests of which yield a great and diverse range of products. In many areas, the growth of revenues from NTFPs has been faster than the revenue increase from timber (Tewari and Campbell, 1995).

NTFP COLLECTION OUTSIDE THE UNITED STATES

While the majority of this volume deals with NTFPs in two U.S. regions, this paper focuses on issues related to non-timber forest products outside the United States. The tropics are the main focus, as much of the NTFP literature deals with forests in these areas. While the rural poor in parts of the United States (Appalachia for example), and similar less developed regions of Canada and Western Europe may be as dependant on NTFPs as citizens of some poor tropical countries, it is likely that the scale differs. Some generalizations are made below that distinguish the collection and marketing of NTFPs in the tropics from that in the United States.

People living in and around forests in the developing world tend to be economically depressed and labor is an order of magnitude cheaper than in the United States.[1] Unemployment is often high and governments do not provide unemployment benefits to rural people. Additionally, indigenous people living around forests often have strong cultural and religious links to the forest that date back several generations. While this might also be true of Native Americans and other long-time residents of forested areas, the United States is largely comprised of recent settlers with no such attachments.

In many parts of the world, traditional medicines, many of which are derived from forests (Balick, 1990; Balick and Mendelsohn, 1992), are the basis of much of the primary health care (Farnsworth et al. 1985). In these mostly tropical nations, western health care is literally beyond the reach of many rural people due to the high costs of time and money it takes to travel to urban centers where such treatments are available. Also, as traditional healers usually know their patients and can provide greater reassurance than busy hospitals, they are in a better position than hospitals to provide care where psychological and psychosomatic complaints occur (Young et al. 1988, DeBeer and McDermott, 1989).

Most homes in the United States have access to motor roads. Many who live in the developing world live a several hour walk from roadhead, and this restricts trade with urban markets. This lack of access to roads and markets restricts the formal market value of many NTFPs, but may encourage greater local consumption of forest resources. Thus, money and mobility are two important factors that separate citizens of the United States and their use of NTFPs from those in many other parts of the world.

NTFPS AROUND THE WORLD-SOME EXAMPLES

An article of this length can only skim over some of the pertinent information on NTFPs. In India alone, over 3,000 plant species produce economically significant products and their use and trade is integral to local economies and cultures (Tewari and Campbell, 1995). In this section I review some of the major types of non-timber forest products and site some relevant examples.

Biomass-based NTFPs: Fuelwood and fodder are low-value products and superior commercial substitutes are available. However, because these products are often available for free, and are ubiquitous, they are among the most important NTFPs. In terms of volume, fuelwood is the most important NTFP to be extracted from forests. It accounts for 94% of the NTFPs taken from the forests of Tamil Nadu, India (Appasamy, 1993) and is the predominant NTFP in many parts of Asia, Africa, and the dry tropics around the world. Fuelwood is the major source of energy in large parts of the developing world, and much of this is extracted from the natural forest.[2] In the central Himalaya, for example, fuelwood from the forests is the primary source of energy in over 90% of rural households, and forest-derived green fodder takes care of most of the nutritive needs of cattle and goats. In addition leaf litter swept off the forest floor is used to produce a compost fertilizer that is often the only nutrient input in hill agriculture (Thadani, 1999).

Rattans and bamboos: While Rattans are climbing palms (sub-family Calamoideae) and bamboos are grasses (sub-family Bambusiodeae), these are discussed together due to their similarity of structure and uses. Both are used to make light attractive furniture or for weaving baskets and other handicrafts. Economically, rattans and bamboos are the most important NTFPs in Asia. The trade in finished rattan products in 1988 was estimated at US $2.7 billion (de Beer and McDermott, 1989). As rattan requires arboreal support and shade, its cultivation does not require the clear felling of existing forest. Yet, the cultivation of rattan can be about eight times more profitable than rain-fed rice (Godoy and Feaw, 1989). Bamboo, in addition to its structural uses, is also used in paper manufacture and its shoots are an important food in southeast Asia.

Edible fruit, nuts and other foods: Edible NTFPs not only have considerable economic values in some areas, but also provide food security to low income populations during drought and famines (FAO, 1989). Many of the more important edible fruits of the world are no longer regarded as NTFP crops as they are extensively cultivated in commercial plantations. Even regionally important fruits such as mangosteen (*Garcinia mangostana* L.) and durian (*Durio zibethinus* Murray), important in southeast Asia, are increasingly obtained through cultivation, although several varieties of durian are still harvested from forests. Among nuts, the Brazil nut (*Bertholletia excelsa* H.B.K.) is extracted mainly from wild trees in the forests of Amazonia (Smith et al. 1992) and rubber collectors often derive a significant proportion of their income from this species. The open capsules of the Brazil nut are also sometimes used to collect latex from the rubber tree. Indigenous people throughout the world eat diverse plant parts. For example, Sago, a starchy extract from the pit of Southeast Asian palms and cycads, is a locally important food (de Beer and McDermott, 1996). The sap tapped from the inflorescence of several other palms, such as *Nypa fruticans* Wurmb. and *Caryota urens* L. is used to make sugar and 'toddy,' an alcoholic beverage.

Medicinal plants: Many modern medicines originated in forests around the world. Some, such as quinine–extracted from the bark of *Cinchona officinalis* L.–and common aspirin–first extracted from the willow–are now manufactured synthetically. Others, such as the bark of the African cherry (*Prunus africana* Hook.f. Kalkman)–the major source of the extract used to cure benign prostatic hyperplasia–are extracted almost exclusively from wild forest trees (Cunningham and Mbenkum, 1993). In addition, there exist thousands of medicinal plants administered locally by shamans, medicine men and community elders. Many of these will become commercially important as western medicine rediscovers what indigenous people often already know.

Resins and latex: Rubber, extracted from *Hevea brasiliensis* (Willd. ExA. Juss.) Mull. Arg. in Amazonian forests, is the best known NTFP in this group. The extractive reserves described in a later section have been formed around rubber tappers in Brazil. Elsewhere, locally available substitutes for *Hevea* rubber are extracted from several other species such as Indian rubber (*Ficus elastica* cv.) and manicoba rubber (*Manihot glaziovii* Mull. Arg.). Pines, both temperate and tropical, are an important source of resin. For example, a thriving industry based around turpentine oil was started by the British in the Indian Himalaya using resin from chir pine (*Pinus roxburghii* Sarg.) and continues to this day despite the availability of synthetic substitutes. Gaharu, the resin from infected heartwood of *Aquilaria* spp. (Thymeliaceae) is an extremely expensive commodity in southeast Asia and is used for making perfumes. Chicle, used in chewing gums and glue, is made from the latex from *Manilkara zapota* (L.) P. Royen trees (Sapotaceae).

Wildlife and wildlife products: The value of wildlife is often ignored in NTFP valuation studies. However, bushmeat is an important source of protein in many parts of the developing world. The meat of wildboar and various deer such as the Sambar (*Cervus unicolor*) and barking deer (*Muntiacus* spp.) is prized throughout much of Asia. In addition, animal products such as honey and bird and turtle eggs are important animal products collected from the forest. Fish from floodplains and lakes are an important component of the diet of some groups. In South Asia, tens of thousands of snake-charmers, monkey and bear performers make their living by exhibiting these animals. Hunters and trappers derive valuable skins and meat from crocodiles and pythons in the tropics, and minks and foxes in temperate regions. Several birds are prized for their plumes while others are trapped for the pet trade. The tusks of elephants, used to make personal seals and decorative items, the horns of rhinos, and the bones of tigers used in aphrodisiacs and various oriental medicines are all valuable commodities in Asia. Often they are worth their weight in gold.

Cultural, religious and aesthetic commodities: Because economic valuation of these products is difficult to carry out, they have remained understudied. Orchids and other exotic ornamentals plucked from forests, feathers and bones used as ornaments, and other such apparently 'worthless' products are greatly prized by people around the world. These products, while not essential to the physical survival of people, go a long way toward maintaining cultural traditions and improving quality of life. A case in point are Xate, the fronds from *Chamaeodorea* spp. (Palms–Arecaceae), which are used in floral arrangements for weddings. In addition to their immediate cultural, religious, and aesthetic worth, many such products also have economic values. Thus, while orchids may be used for decorating the homes of harvesters, they many also fetch high prices in commerce.

PROMISES OF SUSTAINABLE PROSPERITY

Several researchers published studies on the valuation of tropical forests in the late 1980's and early 1990's. The popularity and perceived need for such studies greatly increased after Peters et al. (1989) published their valuation of an Amazonian rainforest. These authors found that the net present value (NPV) of a tract of forest in Peru was highest if the forest was used for sustainable extraction of fruits and rubber latex, as opposed to being logged for timber and converted to pasture. They proposed that the exploitation of non-wood forest resources was the most "profitable method for integrating the use and conservation of Amazonian forests" (Peters et al. 1989, p. 656). In an analysis of two plots from secondary hardwood forests in Belize, Balick and Mendelsohn (1992) showed that the value of medicinal plants alone could justify the protection of some areas of rainforest as extractive reserves.

This line of research contended that developing marketable NTFPs creates an economic justification for the preservation of forests that might otherwise be destroyed by developmental pressures. While the removal of timber often results in the removal or destruction of all other components of the forest, it was argued that the removal of non-timber products usually allows the structure of the forest to be preserved (Godoy and Feaw, 1989). This maintains most of the intangible environmental benefits, which may exceed the worth of the economic benefits from a forest. Also, forest policies that ignore NTFPs often anger local people, leading villagers to vent their disapproval through burning of forests and plantations (Gunatilleke et al. 1993).

NTFPs were also believed to lead to a more equitable distribution of wealth as the poor benefit most from NTFPs, while the rich benefit more from timber extraction. Also, women play an important role in NTFP trade. In the central Himalaya, fuelwood and fodder collection is largely carried out by women and Ganeshan (1993) found similar conditions in the Mudumalai forests in south India. In the humid forests of Cameroon, Ndoye et al. (1998) estimate that 94% of NTFP traders are women. It was thought that the control of NTFP activities by women could lead to a more equitable distribution of forestry benefits within communities (Tewari and Campbell, 1995). Thus, in NTFPs conservationists found an "alluring mix of ecological, economic and social justifications for preserving rain-forest lands in a relatively pristine condition" (Salafsky et al. 1993).

EXTRACTIVE RESERVES

A movement started by Brazilian rubber tappers, or caboclos, in the late 1970's gradually drew the support of Western scientists. This proposal for the creation of 'extractive reserves,' where social groups that historically occupy lands would be provided legal rights to utilize forest products in an ecologically sustainable fashion, offered hope for the protection of rain forest areas (Allegretti, 1990). Families based in extractive reserves produce a mix of subsistence and market goods, combining commercial production of broadly occurring species such as rubber (*Hevea brasiliensis*) and Brazil nut (*Bertholletia excelsa*) with direct consumption of other forest products that occur in more restricted distributions. Shifting cultivation and gathering of native plants (such as fruits and palm hearts of the 'acai' palm [*Euterpe oleracea* Mart.] as well as various medicinal plants), fishing, and hunting help supplement their subsistence lifestyle (Allegretti, 1990). In a study of caboclos in the Amazon estuary Anderson (1990b) found that their use of the forest was not merely extractive, as had been previously thought, but involved management practices that favored desirable species while less desirable competitors were eliminated or thinned. He concluded that the "essential ingredient for reconciling extraction and forest management is a rural population with both a knowledge of, and respect for, forest resources."

While extractive reserves appeared to be a promising development strategy, a closer examination revealed several problems inherent in such systems and the early enthusiasm was soon tempered by evidence of the limitations of extractive reserves (Salafsky et al. 1993). NTFP collection rarely seems to provide more than a marginal existence for people and it never appears to alleviate their poverty. People dependent on NTFP collection have limited access to health care, education, or surplus income. Economic hardships often force these people to maximize short-term returns, frequently at the cost of long-term sustainability. Thus, while concepts such as extractive reserves hold much charm, in reality they may be prone to environmental degradation and a marginal existence for NTFP extractors. Such reserves are also usually viable only where population densities are very low. For example, rubber trappers in Brazil require 300-500 ha per family to collect sufficient rubber, Brazil nuts and other NTFPs (Fearnside, 1989). In the heavily populated countries of south and southeast Asia, such large tracts of land are simply not available and efforts to create them would lead to inequitable distribution of forest resources. Also, extractive reserves may be unable to compete with plantations or agroforestry systems that produce similar products through more intensive land use (Salafsky et al. 1993).

NTFP VALUATION

Many proponents of non-timber forest products believed that economic valuation of NTFPs was critical if resource managers and policymakers were to decide between alternate uses of forests. However, the commercial value of an NTFP is difficult to calculate. Several NTFPs, while highly valued by people who use them, are of little commercial value until western science 'discovers' them. An example would be that of the neem tree (*Azadirachta indica* A. Juss.), which has been used by villagers in India for several centuries as an antiseptic, anti-fungal, insecticide, and for its properties to cure ailments such as diarrhea and dysentery (Rawat, 1995). However, it had limited commercial value until Western multinational companies became aware of the species and patented an improved version of the naturally occurring pesticide. As a result, the species gained tremendous economic and political importance (Shayana, 1998).

As extraction rates of most NTFPs are not well documented, and since many are bartered and never enter the commercial market, economic valuation is very labor intensive. Many of the values of these products are social rather than economic and factoring these into calculations remains elusive. The differences in methodologies applied by researchers has also led to very significant differences in estimates of the values of NTFPs (Godoy and Lubowski, 1992; Godoy et al. 1993, Chopra, 1993).

Uncertainties of price and tenure are another drawback that make economic valuations uncertain. Cheap commercial substitutes can destroy NTFP markets. The traditional umbrella and clog industry in Indonesia has, for example, given way to plastic substitutes (Tewari and Campbell, 1995). In addition, diminishing supplies of NTFPs, limited access to financial institutions, and problems in income sharing between NTFP suppliers and intermediaries are constraints faced by small enterprises (Tewari and Campbell, 1995).

Further, governmental restrictions on commercially important NTFPs may lead to destruction of the market for an NTFP. The commercialization of medicinal herbs in the Himalaya resulted in an increase in the collection period of these herbs by the Bhotiya, an agro-pastoral community, from two to five months. As the economic returns increased as a result of price increase, the Bhotiya, who previously integrated the collection of herbs from alpine meadows with livestock grazing in these areas, decoupled these two activities and herb collection became a full time occupation for some (Farooquee and Saxena, 1996). As contractors and middlemen got involved in the trade, the Bhotiya pastoralists became mere laborers involved in this trade, levels of collection became unsustainable, and this finally resulted in a government-imposed ban on collection and trade in 1992 (Farooquee and Saxena, 1996).

LIMITATIONS OF ECONOMIC VALUATIONS

While the last decade has seen an abundance of papers on tropical NTFPs, an inordinately large number appear to try to justify the importance of NTFPs based on their economic worth. Often ignored, are the values that cannot be quantified. Fruits, rich in vitamins and nutrients, may be essential in preventing certain malnutrition-related diseases (e.g., see Caldwell, 1972). While the fruits may not have any commercial value, their benefit to health is hard to quantify in dollar terms. Similarly, a valuation study cannot attach a commercial value to a trophy–a tiger tooth, for example–that a forest dweller has acquired at great peril to his life, and which might be his most valued possession. Attractive varieties of *Coleus* and *Heliconia* are transplanted from forests to gardens in villages in New Guinea (deBeer and McDermott, 1989). Such NTFPs, gathered for aesthetic purposes, may be used locally and have no commercial value but do much to improve the quality of life of the consumer.

Further, while studies usually focus on remote indigenous groups who are dependent on NTFPs as a primary source of income, a far greater number of people are partially dependent on NTFPs. NTFP collection and processing is often a secondary occupation for farmers and other rural people, who may be involved in these activities only during slack periods of the agricultural cycle and derive only a small part of their household income from NTFP-related activities. Fuelwood collection in the Himalaya peaks after the agricultural season, when few sources of income are available. In many parts of the tropics there exists an integrated system of resource utilization where hunting and gathering of forest products is integrated with the farming systems. In the central Himalaya, for example the agricultural system is dependent on inputs from the forest (Singh and Singh, 1992). In a region where the topography precludes the use of heavy commercial fertilizers and mechanical farming, the input of compost fertilizer made from forest leaf litter is essential to maintain soil fertility. Energy is mostly derived from forest fuelwood, fields are ploughed by drought animals fed on forest fodder, traditional medicine systems rely on herbs from the forest, and a variety of socio-cultural norms integrate this region closely with the forest. Although relatively small, the NTFP collected or income earned may be an essential supplement. Yet economic valuations rarely capture these contributions to the livelihoods of the individuals and communities that depend on them.

Social, cultural and religious values of NTFPs also cannot be evaluated and remain inadequately studied. For example, the Batak hunter-gatherers of Palawan, Philippines, spend two weeks every March in small camps situated in special forest sites (see deBeer and McDermott, 1996). Their exclusive objective is to carry out the 'Lambai bee ritual,' during which the spirits of *Apis dorsata* bees are honored with prayers, dances and songs. Such rituals and festivities, often connected with NTFPs, are likely to be instrumental for maintaining social structures within communities.

COMMERCIALIZATION OF NTFPS

Commercializing non-timber forest products may lead to a variety of problems. Perhaps the greatest concern has been the depletion of NTFPs that have a high commercial value attached to them. The large-scale commercial exploitation of NTFPs can lead to their over-exploitation and consequent depletion (Soemarwoto, 1990) or the increased availability may cause a crash in prices (Balick and Mendelsohn, 1992). In the Brazilian Amazon, the decline in populations of Brazil nut trees and the extirpation of tapirs by hunting has been observed (Nepstad et al. 1992). The palm-heart canning industry resulted in the destruction of natural stands of *Euterpe edulis* Mart. in southern Brazil. Similarly, the depletion of rosewood trees (*Aniba rosaeodora* Ducke), the resource base for rosewood oil in Amazonas, resulted in reduction of the number of processing plants from 103 in 1960 to 20 in 1986 (Richards, 1993). Cunningham and Mbenkum (1993) report the over-exploitation of *Prunus africana* in the Afro-montane forests of Cameroon after 1966, when its bark extract was patented. The high value of elephant tusks and rhino horns has endangered and caused local extinction's of populations of these large animals in African and Asian forests.

Thus, while biotic impoverishment is substantially higher with other types of land use such as timber harvest and cattle ranching, a depletion of commercially important species may often accompany

NTFP extraction. Initiatives that promote participatory methods to manage NTFPs have the best chance of success. Not only does this allow NTFP gatherers to continue with their trade but it has the potential of incorporating conservation norms among people. However, such methods are hard to administer and require skilled personnel and a flexibility of operation that governments are usually not comfortable with. Plantations of NTFPs, best exemplified by rubber and cocoa plantations, offer hope of commercially exploiting at least some NTFPs. But these, too, are criticized for resulting in lowered biodiversity and other problems usually associated with monocultures. Also, products harvested from plantations are not thought of as NTFPs (de Beer and McDermott, 1989).

SOME RESEARCH AND POLICY ISSUES

Some of the major issues relating to the extraction and trade of non-timber forest products are listed below:

- *Tenure:* NTFP collectors often use Government lands on which they do not possess clearly defined legal rights. This makes these people very vulnerable to the whims of politicians and officials, and insecure about their future. The absence of clear tenure leads to a lack of incentives for NTFP harvesters to conserve available resources for long-term use.
- *Governmental interventions:* NTFP collection is often the domain of the poorest of a population. Tribals and unemployed rural poor are most commonly employed in NTFP collection. These people have neither the knowledge nor access to, improved processing and packaging technologies. Usually, they sell unprocessed produce to middlemen who often monopolize the trade and take most of the profits (Poffenberger et al. 1990). Processing these NTFPs can add considerable value to the raw product. With altruistic objectives, the governments of several countries have gotten involved in NTFP collection and trade. Frequently this has had disastrous consequences for the collectors, who cannot cope with the bureaucracy and payment delays. In India, the nationalization of several NTFPs has had adverse effects for tribals and rural poor and resulted in the sharp decline of several NTFPs following their nationalization (Chambers et al.

1990, p. 149). The banning of rattan exports from Borneo, Indonesia depressed the price of raw rattan by 40%. As a result smallholder palm cultivators switched to alternate crops and suffered income losses (Safran and Godoy, 1992).

- *Temporal availability of products:* Ecological constraints such as fruiting, flowering or leafing phenology, physical constraints such as flooded soils that prevent harvest, or economic constraints such as seasonal demand can dictate that a product only be periodically harvestable. Workers specializing in NTFP collection therefore need to gather a mix of products whose availability and demand periods are staggered in a way that generates employment throughout the year. In Petén, for example, the three main non-timber forest products, chicle latex, xate fronds and allspice fruits are harvested at different times of the year in a sequential manner that allows harvesters to earn steady cash income throughout the year (Salafsky et al., 1993).
- *Changes in ecosystem function and stability:* Extraction of forest-products can change the species composition, alter fire regimes, and facilitate the invasion of invasive species. For example, the collection of fuelwood and fodder from the forests of the central Himalaya leads to a replacement of banj oak by chir pine, while the tapping of chir pine trees for resin makes them susceptible to ground fires (Thadani, 1999).
- *Nutrient depletion:* Fruit are often rich in nutrients and, while these comprise only a small amount of the forest biomass, it is possible that sufficient nutrients are removed to cause deficiencies of critical nutrients. Tropical forests often grow on very poor soils and efficient nutrient recycling mechanisms help maintain ecosystem stability (Jordon, 1985). Leaf removal for compost production, for example, can result in lower levels of nutrients in the oak stands of the central Himalaya (Thadani, 1999). Thus studies looking at sustainability issues need to focus on ecosystem-level issues such as nutrient removal in addition to studies of the population structures of important plants.
- *Large land requirements:* Extractive reserves can generally support only low human population densities (Salafsky et al. 1993). In the Amazon, for example, Brazil nut harvesters use 300-500 ha of forests for each family (Fearnside, 1989). In countries that

have very high population densities, it may be hard to justify preserving such large areas for relatively few families.

- *Multiple use management:* Integrating the harvest and management of NTFPs with forest management for timber has been suggested as the most effective means to manage tropical forests and one that is key to economic profitability (Panayotou and Ashton, 1992; Salick et al. 1995). There is a long history of this kind of integrated extraction of timber and NTFPs (Whitmore, 1990).
- *Markets:* The marketing and processing of NTFPs has not received due attention. Unlike tropical timber, which is sold in international markets and generates substantial foreign exchange, NTFPs are sold largely in local markets through decentralized trade networks that involve large numbers of forest collectors, middlemen and shop-owners. Such networks are hard to monitor and get ignored in national accounting schemes, which leads to an under-valuation of these products (Peters et al. 1989; Appasamy, 1993). However, this situation may gradually be changing in at least some parts of the world.

NOTES

1. The collection of NTFPs is labor intensive. As wages rise, collection of many NTFPs becomes economically unviable. In Asia and Africa, where labor costs may be an order of magnitude less than in the United States, a much broader spectrum of forest products can be collected. For example, the Soliga, an aboriginal tribe of southwest India spend over half their time in NTFP collection, from which they earn only about $250 annually or less than $1 per day (Hegde et al. 1996).

2. Overexploitation of this resource has led to severe shortages (Pimental et al. 1986) and women, who are the primary gatherers of fuelwood, have to spend ever increasing time searching for this resource. This has led to a decline in time spent in the field, and a reduction in crop yields is speculated as a consequence (Eckholm, 1975). According to some estimates, 2.4 billion people, many in India and Africa, may suffer fuelwood shortages by the year 2000 (see Smith et al. 1992).

REFERENCES

Allegretti, M.H. 1990. Extractive reserves: an alternative for reconciling development and environmental conservation in Amazonia. In: pp. 252-264. A.B. Anderson (ed.). Alternatives to deforestation: steps towards sustainable use of the Amazon rain forest. Columbia University Press, New York.

Anderson, A.B. 1990a. Deforestation in Amazonia: dynamics, causes and alternatives. In: pp. 1-23. A.B. Anderson (ed.). Alternatives to deforestation: steps towards sustainable use of the Amazon rain forest. Columbia University Press, New York.

Anderson, A.B. 1990b. Extraction and forest management by rural inhabitants in the Amazon estuary. In: pp. 65-85. A.B. Anderson (ed.). Alternatives to deforestation: steps towards sustainable use of the Amazon rain forest. Columbia University Press, New York.

Appasamy, P.P. 1993. Role of non-timber forest products in a subsistence economy: the case of a joint forestry project in India. Economic Botany 47(3): 258-267.

Balick, M.J. 1990. Ethnobotany and the identification of therapeutic agents from the rainforests. In: D.J. Chadwick and J. Marsh (eds). Bioactive compounds from plants. J. Wiley and Sons, Chichester, England.

Balick, M.J. and R. Mendelsohn. 1992. Assessing the economic value of traditional medicines from tropical rain forests. Conservation Biology 6(1):128-130.

Caldwell, M.J. and I.C. Enoch. 1972. Riboflavin content of Malaysian leaf vegetables. Ecology of Food and Nutrition 1:301-312.

Chambers, R., N.C. Saxena and S. Tushaar. 1990. To the hands of the poor: water and trees. Oxford & IBH, New Delhi, India. 273 pp.

Chopra, K. 1993. The value of non-timber forest products: an estimation for tropical deciduous forests in India. Economic Botany 47(3):251-257.

Cunningham, A.B. and F.T. Mbenkum. 1993. Sustainability of harvesting *Prunus* African bark in Cameroon: a medicinal plant in international trade. People and Plant. Working Paper 2, May 1993. UNESCO, Paris.

De Beer, J.H. and M.J. McDermott. 1989. The economic value of non-timber forest products in Southeast Asia. Netherlands committee for IUCN. Amsterdam. 175p.

De Beer, J.H. and M.J. McDermott. 1996. The economic value of non-timber forest products in Southeast Asia, 2nd edition. Netherlands committee for IUCN. Amsterdam. 197p.

Eckholm, E. 1975. The other energy crisis: firewood. Paper 1. Worldwatch Institute, Washington DC.

Farnsworth, N.L., O. Akerele, A.S. Bingel, D.D. Soejarto and Z.G. Guo. 1985. Medicinal plants in therapy. Bull. WHO 63:965-981.

Farooquee, N.A. and K.G. Saxena. 1996. Conservation and utilization of medicinal plants in high hills of the central Himalayas. Environmental Conservation 23(1): 75-80.

Fearnside, P.M. 1989. Extractive reserves in Brazilian Amazonia: an opportunity to maintain tropical rain forest under sustainable use. BioScience 39(6):387-393.

FAO. 1989. Forestry and food security. Food and Agriculture Organization, Rome, Italy. 128p.

Freezailah, B.C.Y. 1992. Foreword. In: T. Panayotou and P.S. Ashton. Not by timber alone: economics and ecology for sustaining tropical forests. Island Press, Washington DC. 282p.

Ganesan, B. 1993. Extraction of non-timber forest products, including fodder and fuelwood. Economic Botany 47(3):268-274.

Godoy, R. and R. Lubowski. 1992. Guidelines for the economic valuation of non-timber forest products. Current Anthropology 33(4):423-433.

Godoy, R., R. Lubowski and A. Markandya. 1993. A method for the economic valuation of non-timber tropical forest products. Economic Botany 47(3):220-233.

Godoy, R.A. and T.C. Feaw. 1989. The profitability of smallholder rattan cultivation in Southern Borneo, Indonesia. Human Ecology 17(3):347-363.

Gunatilleke, I.A.U.N., C.V.S. Gunatilleke and P. Abeygunawardena. 1993. Interdisciplinary research towards management of non-timber forest resources in lowland rain forests of Sri Lanka. Economic Botany 47(3):282-290.

Hegde, R.S. Suryaprakash, L. Achoth and K.S. Bawa. 1996. Extraction of non-timber forest products in the forests of Bilgiri Rangan Hills, India. 1. Contribution to rural income. Economic Botany 50(3):243-251.

Jordon, C.F. 1985. Nutrient cycling in tropical forest ecosystems. John Wiley and Sons, Chichester, England.

Lanly, J.P. 1982. Tropical forest resources. FAO Forestry Paper No. 30. Food and Agriculture Organization, Rome, Italy. 106p.

Ndoye, O., M.R. Perez and A. Eyebe. 1998. The markets of non-timber forest products in the humid forest zone of Cameroon. Rural Development Forestry Network, Overseas Development Institute, London, England.

Nepstad, D.C., I.F. Brown, L. Luz, A. Alechandre and V. Viana. 1992. Biotic impoverishment of Amazonian forests by rubber tappers, loggers and cattle ranchers. Advances in Economic Botany 9:1-14.

Panayotou, T. and P.S. Ashton. 1992. Not by timber alone: economics and ecology for sustaining tropical forests. Island Press, Washington, DC. 282p.

Peters, C.M., A.H. Gentry and R.O. Mendelsohn. 1989. Valuation of an Amazonian rainforest. Nature 339:655-656.

Pimental, D., W. Dazhong, S. Eigenbrode, H. Lang, D. Emerson and M. Karasik. 1986. Deforestation: interdependency of fuelwood and agriculture. Oikos 46:404-412.

Poffenberger, M. 1996. Non-timber tree products and tenure in India: considerations for future research. In: pp. 70-84. M.P. Shiva and R.B. Mathur (eds.). Management of minor forest products for sustainability. Oxford and IBH Publishing Co., New Delhi, India.

Poffenberger, M., B. McGean and K. Bhatia (eds.). 1990. Forest management partnerships: regenerating India's forests. Executive summary of the workshop on sustainable forestry, Sept 10-12, New Delhi, India.

Poore, D. Conclusions. 1989. In: D. Poore, P. Burgess, J. Palmer, S. Rietbergen and T. Synnott (eds.) No timber without trees: sustainability in the tropical forest. A study for ITTO. Earthscan. London, UK. 252p.

Rawat, G.S. 1995. Neem (*Azadirachta indica*)–nature's drugstore. Indian Forester 121(11):977-980.

Richards, M. 1993. The potential of non-timber forest products in sustainable natural forest management in Amazonia. Commonwealth Forestry Review 72(1):21-27.

Safran, E.B. and R.A. Godoy. 1992/3. Effects of government policies on smallholder palm cultivation. Human Organisation.

Salafsky, N., B.L. Dugelby and J.W. Terborgh. 1993. Can extractive reserves save the rain forest? An ecological and socioeconomic comparison of nontimber forest product extraction systems in Petén, Guatemala and West Kalimantan, Indonesia. Conservation Biology 7(1):39-52.

Salick, J., A. Mejia and T. Anderson. 1995. Non-timber forest products integrated with natural forest management, Rio San Juan, Nicaragua. Ecological Applications 5(4):878-895.

Shayana, K. 1998. United States patent prior art rules and the neem controversy: a case of subject-matter imperialism? Biodiversity and Conservation 7(1):27-39.

Singh, J.S. and S.P. Singh. 1992. Forests of Himalaya. Gyanodaya Prakashan, Nainital. 294p.

Smith, N.J.H., J.T. Williams, D.L. Plucknett and J.P. Talbot. 1992. Tropical forests and their crops. Cornell University Press, Ithaca, NY. 568p.

Soemarwoto, O. 1990. Forestry and non-wood products: a developing country's perspective. In: S. Counsell and T. Rice (eds.). The rainforest harvest: sustainable strategies for saving the tropical forests. Friends of the Earth Trust Ltd., London, England. pp. 64-69.

Tewari, D.D. 1998. Economics and management of non-timber forest products: a case study of Gujarat, India. Oxford and IBH Publishing Co. Pvt. Ltd., New Delhi. 164p.

Tewari, D.D. and J.Y. Campbell. 1995. Developing and sustaining non-timber forest products: some policy issues and concerns with special reference to India. Journal of Sustainable Forestry 3(1):53-79.

Thadani, R. 1999. Disturbance, microclimate and the competitive dynamics of tree seedlings in banj oak (*Quercus leucotrichophora*) forests of the central Himalaya, India. Ph.D. thesis, Yale University, New Haven, CT. 318p.

Whitmore, T.C. 1990. An introduction to tropical rainforests. Clarendon, Oxford, England.

World Resources Institute. 1990. The World Bank in the Forest Sector: A Global Policy Paper.

Young, D., G. Ingram and I. Swartz. 1988. The persistence of traditional medicine in the modern world. Cultural Survival Quarterly 12(1) 39-41.

Brief Overview of Historical Non-Timber Forest Product Use in the U.S. Pacific Northwest and Upper Midwest

Marla Emery
Shandra L. O'Halek

SUMMARY. Non-timber forest products (NTFPs) have sustained indigenous and immigrant populations alike since their arrival in North America. This brief overview focuses on the historical use of NTFPs in the U.S. Pacific Northwest and Upper Midwest. Drawing on sources as diverse as accounts by early European arrivals, archaeological evidence, and contemporary ethnobotanical studies, we touch on documented uses of forest vegetation from prehistory to the present century. The residents of these regions have used NTFPs for food, medicine, and cultural materials. NTFPs have met their livelihood needs through subsistence uses and both non-market and market exchanges. We conclude

Marla R. Emery is Research Geographer, USDA Forest Service, Northeastern Research Station, 705 Spear Street, P.O. Box 968, Burlington, VT 05402-0968 USA (E-mail: memery@fs.fed.us).

Shandra L. O'Halek is Senior Planner for Mason County, WA.

The authors are indebted to Susan Alexander and David Pilz, USDA Forest Service Pacific Northwest Research Station, for their assistance in identifying scientific names for NTFP species in that region. And to Beth Lynch, formerly with the Great Lakes Indian Fish and Wildlife Commission, Dana Richter, Michigan Technological University, and Jan Schultz, Hiawatha National Forest, for sharing their knowledge of the proper nomenclature for Upper Midwestern species.

[Haworth co-indexing entry note]: "Brief Overview of Historical Non-Timber Forest Product Use in the U.S. Pacific Northwest and Upper Midwest." Emery, Marla, and Shandra L. O'Halek. Co-published simultaneously in *Journal of Sustainable Forestry* (Food Products Press, an imprint of The Haworth Press, Inc.) Vol. 13, No. 3/4, 2001, pp. 25-30; and: *Non-Timber Forest Products: Medicinal Herbs, Fungi, Edible Fruits and Nuts, and Other Natural Products from the Forest* (ed: Marla R. Emery, and Rebecca J. McLain) Food Products Press, an imprint of The Haworth Press, Inc., 2001, pp. 25-30. Single or multiple copies of this article are available for a fee from The Haworth Document Delivery Service [1-800-342-9678, 9:00 a.m. - 5:00 p.m. (EST). E-mail address: getinfo@haworthpressinc.com].

that in spite of U.S incorporation into a global market-based economy, there is notable continuity in the harvest and use of NTFPs in the United States from prehistory to current times. *[Article copies available for a fee from The Haworth Document Delivery Service: 1-800-342-9678. E-mail address: <getinfo@haworthpressinc.com> Website: <http://www.HaworthPress.com>]*

KEYWORDS. Non-timber forest products, environmental history, human-forest interactions, Michigan, Pacific Northwest

Non-timber forest products (NTFPs) have been important to the livelihoods of the inhabitants of North America from prehistoric times to the present. As elsewhere in the world, early inhabitants of the forested portions of North America made extensive use of the vegetation that surrounded them. Archaeological evidence indicates, for example, that by 6,000 B.C. Native American residents of the Upper Great Lakes Basin relied heavily on plant foods gathered in the region's mixed conifer-deciduous forests (Cleland, 1983).

For the last four centuries, NTFPs have continued to help sustain Native and other Americans. Early historical documents record the importance of NTFPs for the human population of North America. Missionaries writing to their European headquarters between 1610 and 1791 described the use of plant matter from the northeastern woods to meet a variety of human needs (Society of Jesus, 1898). Among other items, their reports frequently mentioned the use of birch bark (*Betula papyrifera* Marsh.) for eating utensils and bedding and as a construction material for dwellings and canoes. They also noted the use of unspecified barks for purposes as diverse as musical instruments, funeral pyres, and poultices for wounds and skin lesions. When he drafted his map of New France (northeastern North America) in 1632, Samuel Champlain (1850) recorded what he believed were the most important cultural, economic, and political features of the province. These included a "lieu ou les sauvages font secherie de framboises et blues tous les ans" ("place where indians dry raspberries (*Rubus* spp.) and blueberries (*Vaccinium* spp.) each year." Learning from the indigenous population and, no doubt, from their own need and experience, early colonists also relied heavily on berries, nuts, and other wild edibles for sustenance (Cronon, 1983; Williams, 1989).

In addition to their subsistence uses, NTFPs have long been important in nonmarket exchanges such as barter and gift giving. For example, until the appearance of trading posts in the late 1700's, the indigenous people of the Olympic Peninsula in the Pacific Northwest were considered especially wealthy because of the abundance of land and sea resources available to them. Historical use of 300 plants from this region for food, flavorings, or spiritual purposes has been documented. Berries were gathered and eaten fresh or dried for storage. Red huckleberry (*Vaccinium parvifolium* Smith) was used for fishing bait because the round red berry simulates roe. Roots such as various fern stocks, tiger lily (*Lilium columbianum* Hanson), clover (*Trifolium* spp.), and lupine (*Lupinus* spp.) were surrounded by moist salal (*Gaultheria shallon* Pursh) and sword fern (*Polystichum munticum* [Kaulf.] Presl) and pit-cooked to convert the bitter emuline to a sweetened fructose. Materials for eating utensils, arrows, paddles, paints, dyes, and shampoos were gathered from the forest. Many of these products were traded to other tribes and, in some cases, traveled far from the Olympic Peninsula. Whether for direct consumption or exchange, ongoing harvest of NTFPs was critical and gatherers used sophisticated plant management techniques that included selective harvesting at specific life stages, replanting, pruning, and landscape manipulation. Ceremonies were performed before harvesting to show respect for plant resources and gratitude to the higher powers that provided for human needs. These practices and spiritual traditions in the region safeguarded and sustained the forest for future generations (Turner and Peacock, submitted 2001).

As the United States became integrated into the world economy, NTFPs began to have market as well as nonmarket uses. Commerce in ginseng (*Panax quinquefolius* L.) began in the early 1700's, with roots harvested throughout eastern forests and sold primarily to China. This trade reached a peak of $6 million in 1875. One measure of the importance of NTFPs in the mid-1800s is evident in government statistics. East of the Mississippi River, income from ginseng and "all other forest products"-probably including maple syrup (*Acer saccharum* Marsh.), wild fruits, and honey-was deemed to be a significant enough component of the national economy for enumeration in the 1840 U.S. census (Williams, 1989).

In the Pacific Northwest, the influx of trappers, loggers, and railroads created demand for new products and the region's forest materials assumed a commercial worth. By 1871, the harvesting of conifer seeds for reforestation efforts became an important industry in the Pacific Northwest (Gerdes, 1996). The earliest commercial harvesting of floral greens–which remains the most enduring and stable year-round NTFP industry in the region–can be credited to Sam Roake, an enterprising immigrant from England who saw the usefulness and beauty of the elegant fronds of the Northwest sword fern, also known as "brush" (Heckman, 1951). In 1915, he bought an old horse barn by a railroad in Castle Rock, Washington, and began shipping brush to California. Brush sheds sprang up over night and soon other products such as glossy sprays of huckleberry and large-leafed salal boughs were sought by restaurants, stores, and funeral parlors all over the United States. In the years that followed, plants such as Oregon Grape (*Berberis nervosa Pursh*), bear grass (*Xerophyllum texnas* Nutt.), and scotch broom (*Cytisus scoparius* [L.] Link) moved into the brush sheds for sale to international markets.

Industrialization did not eliminate NTFPs as an important livelihood strategy for North Americans. Frances Densmore (1974) documented the use of wild plants for food, medicine, dyes, crafts, and utensils by Native Americans in the Upper Midwest during the first quarter of the 20th Century. Photographs from that period and region attest to the continued importance of NTFPs for much of the European American population as well. They were particularly critical for many households during the Depression. For example, older residents of Michigan's Upper Peninsula relate stories of how blueberries helped sustain their families and refugees from the region's cities during this difficult period, and of how individuals and entire families spent summers camped in the blueberry fields. From their first ripening to the first freeze, children and adults picked berries from morning until evening (Emery). These berries provided for local households in three ways: they were eaten fresh, dozens of quarts were canned for family consumption during the winter, and berries were sold to buyers who sent them by truck or rail to Chicago and other regional metropolises.

Reliance on NTFPs for food, medicines, and other needs persists to this day. Native and other Americans throughout the country have continued to harvest, use, trade, and sell NTFPs. In the early 1970's, the Foxfire books documented the contemporary ginseng trade and seasonal use of wild plant foods in the southern Appalachians (Wigginton, 1975). Currently, more than 100 NTFPs are gathered in the forests of Michigan's Upper Peninsula (see Emery, this issue). U.S.

forests also continue to provide raw materials for the pharmaceutical industry, as they have for more than 100 years. One example is Cascara sagrada (*Rhamnus purshiana* DC.). Named by monks in California (its common name means "holy bark"), Cascara was first extracted for use in laxatives in 1877 by researchers at Parke-Davis Company (Hinterberger, 1974). Recently, the Pacific yew tree (*Taxus brevifoli* Nutt.) was used by the pharmaceutical company Bristol-Meyers Squibb as a source of taxol, a drug used in ovarian and breast cancer treatment (Vance, 1995).

As this brief overview suggests, NTFPs have sustained indigenous and immigrant populations alike and there is notable continuity in their harvest and use in the United States from prehistory to current times. In the past few decades, the abundance of products in forest understories have been looked to as both regular livelihood sources and as buffers in times of economic difficulty. The remainder of this issue draws on recent research in the Pacific Northwest and the Upper Midwest that explores the social and ecological characteristics of NTFPs in northern North America at the end of the 20th century.

REFERENCES

Cleland, C.E. 1983. Indians in a changing environment. In S.L. Flader (ed.). The Great Lakes forest: an environmental and social history. University of Minnesota Press, Minneapolis.

Cronon, W. 1983. Changes in the land: Indians, colonists, and the ecology of New England. Hill & Wang, New York.

Densmore, F. 1974. How Indians use wild plants for food & crafts (formerly titled Uses of plants by the Chippewa Indians). Dover Publications, Inc., New York.

Emery, M.R. 1998. Invisible livelihoods: Non-timber forest products in Michigan's Upper Peninsula. UMI, Ann Arbor, Michigan.

Heckman, H. 1951. The happy pickers of the high cascades. Saturday Evening Post, October 6, 1951.

Hinterberger, J.J. 1974. A claim to distinction: tons of cascara bark. Seattle Times, June 27, 1974.

Society of Jesus. 1898. Travels and explorations of the Jesuit missionaries in New France, 1610-1791: the original French, Latin, and Italian texts, with English translations and notes; illustrated by portraits, maps, and facsimiles. The Burrows Brothers Company, Cleveland.

Turner, N.J. 1982. Northwest coast root foods. Personal communication.

Turner, N.J. and S. Peacock. Submitted 2001. "The Perennial Paradox": Traditional Plant Management on the Northwest Coast. In Duer, D. and N.J. Turner (eds.).

"Keeping it living": Indigenous plant management on the Northwest Coast. University of Washington Press, Seattle, WA.
Vance, N.C. 1995. Medicinal plants rediscovered. Journal of Forestry 93(3): 8-9.
Wigginton, E. 1975. Foxfire 3. Anchor Books, New York.
Williams, M. 1989. Americans & their forests: A historical geography. Cambridge University Press, Cambridge, UK.

Aboriginal Use of Non-Timber Forest Products in Northwestern North America: Applications and Issues

Nancy J. Turner
Wendy Cocksedge

SUMMARY. Aboriginal peoples in northwestern North America have traditionally used hundreds of different forest plants for food, materials and medicines. Plant products have also been economically important as trading goods. Today there are excellent prospects for aboriginal people to participate in the harvesting and marketing of non-timber forest products, but there are serious issues of access to and control of resources, respect of intellectual property rights, and concerns for conservation of plants and ecosystems that must be addressed. We provide an overview of past, current and potential use of NTFPs by aboriginal peoples in British Columbia and neighboring areas, and discuss the relevant issues and concerns, with recommendations about how these can be accommodated. *[Article copies available for a fee from The Haworth Document Delivery Service: 1-800-342-9678. E-mail address: <getinfo@haworthpressinc.com> Website: <http://www.HaworthPress.com> © 2001 by The Haworth Press, Inc. All rights reserved.]*

Nancy J. Turner is Professor, School of Environmental Studies, University of Victoria, Victoria, B.C. V8W 2Y2 (E-mail: nturner@uvic.ca).

Wendy Cocksedge is a graduate student in the School of Environmental Studies, University of Victoria.

The research was made possible by grants from the Social Sciences and Humanities Research Council (#410-94-1555) and Forest Renewal BC (#HQ96137-RE), to N. T. and colleagues, and from Global Forest Foundation (18-2000-83).

[Haworth co-indexing entry note]: "Aboriginal Use of Non-Timber Forest Products in Northwestern North America: Applications and Issues." Turner, Nancy J., and Wendy Cocksedge. Co-published simultaneously in *Journal of Sustainable Forestry* (Food Products Press, an imprint of The Haworth Press, Inc.) Vol. 13, No. 3/4, 2001, pp. 31-57; and: *Non-Timber Forest Products: Medicinal Herbs, Fungi, Edible Fruits and Nuts, and Other Natural Products from the Forest* (ed: Marla R. Emery, and Rebecca J. McLain) Food Products Press, an imprint of The Haworth Press, Inc., 2001, pp. 31-57. Single or multiple copies of this article are available for a fee from The Haworth Document Delivery Service [1-800-342-9678, 9:00 a.m. - 5:00 p.m. (EST). E-mail address: getinfo@haworthpressinc.com].

KEYWORDS. Traditional food, traditional medicine, indigenous peoples, basketry, British Columbia

INTRODUCTION

In North America, aboriginal peoples, more than any other group, have definite, long-standing vested interests in non-timber forest products (NTFPs). Since time immemorial, they have used a multitude of forest resources to sustain themselves physically, culturally and spiritually. For them, the sustainable use of diverse forest plants for food, materials, and medicines has been, simply put, a way of life. Nor are commercial applications of these resources in any way new; plants and plant products have been traded and exchanged for hundreds, perhaps thousands, of years, both within aboriginal groups and between them (Turner and Loewen, 1998).

There are many products in the marketplace today in North America that were originally derived from, and in some cases, are still produced by aboriginal peoples: wild rice (*Zizania aquatica* L.), maple syrup (*Acer saccharum* Marsh.), black walnuts (*Juglans nigra* L.), pinyon nuts (*Pinus edulis* Engelm.), pecans (*Carya illinoensis* [Wangenh.] K. Koch), wild blueberries (*Vaccinium* spp.), and a number of medicinal products, such as sassafras (*Sassafras albidum* [Nutt.] Nees), goldenseal (*Hydrastis canadensis* L.), sweetflag (*Acorus calamus* L.), and mayapple (*Podophyllum peltatum* L.).[1]

Northwestern North American aboriginal peoples are better known for their connections with the fisheries economy, but have also participated in the trade of botanical forest products, and the potential for their continued participation remains high. Berries, basketry materials and woven art, wild herbs and specialty woods are all traditional products for these peoples. The harvesting of additional products such as floral greens, mushrooms and botanical cosmetics, although they may be relatively new and different, also fits in well to traditional lifeways.

Resource management has always been an issue for indigenous peoples, and this includes all types of forest resources. To maintain and enhance the productivity of their resources, aboriginal peoples have developed many different management strategies, applied at different geographical and ecological scales. Northwestern North American First Peoples, like many other indigenous societies, used fire to maintain a mosaic of different successional stages and habitats to increase diversity and availability of plant resources. They also practiced other techniques such as selective harvesting, pruning, coppicing, tending, weeding, replanting and transplanting of culturally important species. In addition, contrary to colonial views of aboriginal land occupancy, almost all aboriginal cultures in the region have had systems of land tenure and resource ownership and stewardship (Turner, Smith et al., in prep.).

Today, due to acculturation, loss of access to resources and other factors, the lifeways of aboriginal peoples have changed considerably. Many of the plant resources they once harvested in quantity are not used to the same extent at present. Nevertheless most of these plants are still significant culturally, and some, particularly berries, specialty woods and basket materials, are still sought in large quantities. Furthermore, new and different applications for these plants are now contributing to peoples' economic development, and this trend will undoubtedly continue as First Peoples in British Columbia gain increasingly greater access to their traditional lands through the Treaty Making process.

Nevertheless, there are significant problems and issues that must be dealt with if the important aboriginal relationships with NTFPs and the lands that produce them are to be maintained into the future. Aboriginal people are concerned with economic development, with securing and regaining access to their traditional lands, with ensuring that their intellectual property rights and cultural integrity are protected, and, perhaps most importantly, with the conservation of forest ecosystems and their species (see Russo and Etherington, 1999). Agreements for co-management of lands and resources, and for full participation in planning decision-making in resource use are increasing, but too often, aboriginal peoples have been overlooked in these areas (Scientific Panel for Sustainable Forest Practices in Clayoquot Sound, 1995).

In this paper, we first consider some of the traditional non-timber forest products developed and used originally by aboriginal peoples of British Columbia and neighbouring areas. We then discuss recent applications of these traditional resources as well as newly adapted and developed NTFPs relevant to aboriginal peoples. We also examine some of the issues they face in the harvesting and marketing of NTFPs from both aboriginal and non-aboriginal participants. Finally, we stress the necessity and desirability for all resource managers and decision-makers to consult and collaborate with First Peoples as major stakeholders in the NTFP industry.

FIRST PEOPLES AS TRADITIONAL USERS OF NON-TIMBER FOREST PRODUCTS

In all, over 500 plant species are known to have specific cultural applications among aboriginal peoples of northwestern North America, and most of these are forest species. Among individual language groups (e.g., Haida, Kwakwaka'wakw, Nlaka'pamux, Okanagan; see Figure 1), the inventory of named, culturally important plant species ranges from about 120 to 350 species. Many of these are used in more than one way for foods, materials, medicines or spiritual purposes; some have many different applications.

Traditional foods in the region include approximately 135 plant species (and some fungi). Examples of prominent plant foods are provided in Table 1. Table 2 gives examples of plant materials used traditionally in construction and manufacture and as fuels. Examples of medicinal plants and their general traditional applications are provided in Table 3. Many of these species are still widely sought, and still play an important role in peoples' cultural traditions and subsistence.

Within the past century or so, some of these plant resources have taken on a new role in the economy of aboriginal peoples; they have been harvested and sold or traded to the Europeans and other newcomers to the area: these were the original non-timber forest products. In some cases, the quantities harvested for sale have been very large. They included both fresh and processed products, some of which were essential to the survival of early colonists and pioneers. Baskets, for example, were a common trading item, initially among First Peoples, and later to the neighboring settlers; for the latter, new innovative woven products such as teacups and saucers and coffee tables were produced (Turner, 1998). Aboriginal harvesters also sold immense quantities of cranberries (*Vaccinium oxycoccus* L.), huckleberries (*Vaccinium* spp.) and other fruits to the settlers at Fort Victoria, Fort Yale and other trading posts, as well as to the Canadian Pacific Railway for their restaurant fare. Aboriginal people also participated in the harvest of pharmaceutical products such as cascara bark (*Rhamnus purshiana* D.C.), and sold these to local company representatives at Victoria and elsewhere.

FIGURE 1. First Peoples language groups of British Columbia and neighboring areas (prepared by Dawn Loewen and Colin Laroque).

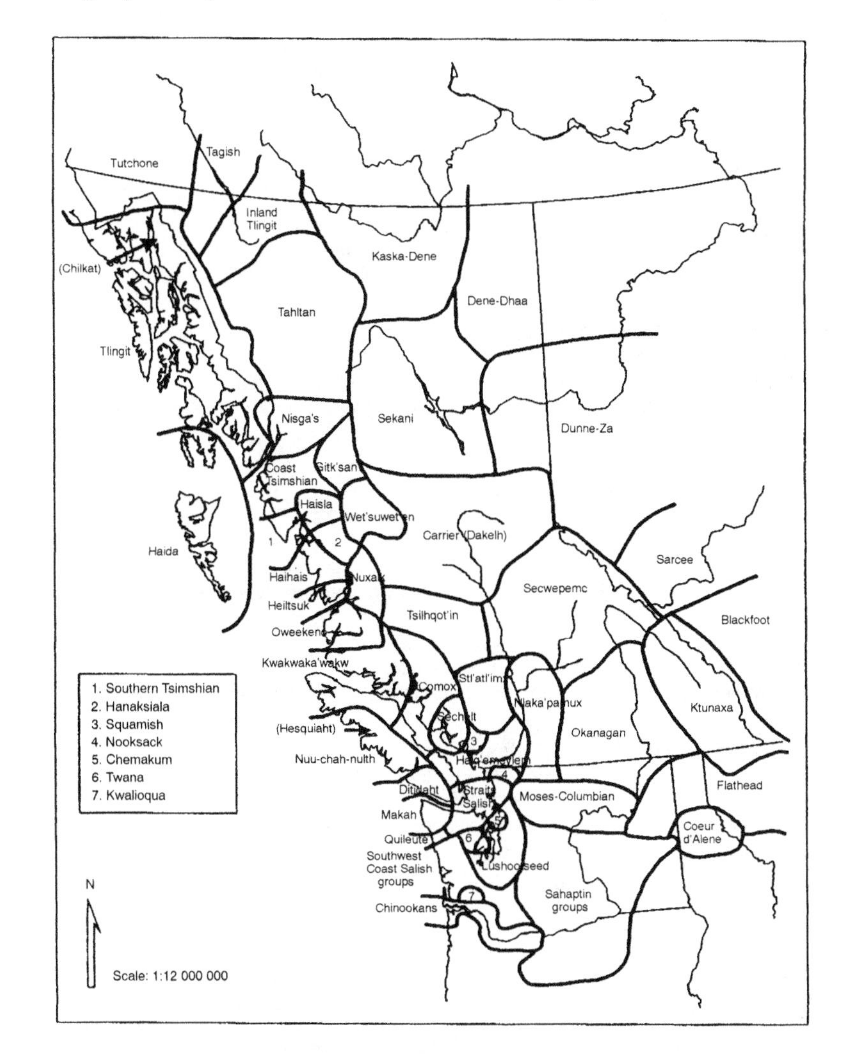

The sale of some of these products continues to the present day. In fact, it could be argued that traditional non-timber forest products and value-added products from aboriginal cultures are more highly valued today than in the past. Table 4 shows selected products and prices of aboriginal art from forest plant materials, as well as of some wild food products produced in whole or in part by aboriginal peoples in B.C.

TABLE 1. Examples of Traditional Plant Foods of Indigenous Peoples of Northwestern North America (Total, about 135 species) (Kuhnlein and Turner, 1991; Turner, 1995, 1997)

Fruits (Total: ~60 species)
Amelanchier alnifolia Nutt; Rosaceae (saskatoon berry)
Fragaria spp. L.; Rosaceae (strawberries)
Gaultheria shallon Pursh.; Ericaceae (salal)
Prunus virginiana L.; Rosaceae (choke cherries)
Pyrus fusca Raf.; Rosaceae (Pacific crabapple)
Rubus spp. L.; Rosaceae (raspberry and related spp.)
Vaccinium spp. L.; Ericaceae (huckleberries, blueberries)
Viburnum spp.; Caprifoliaceae (highbush cranberries)

Green vegetables (Total: ~25 species)
Epilobium angustifolium L.; Onagraceae (fireweed-shoots)
Heracleum lanatum Mitchx.; Apiaceae (cow-parsnip-budstalks, leafstalks)
Lomatium nudicaule (Pursh) Coult. & Roe.; Apiaceae (Indian celery-young leaves and stalks)
Rubus spp.L.; Rosaceae (thimbleberry, salmonberry-shoots)

"Root" vegetables (Total: ~35 species)
Allium cernuum Roth.; Liliaceae (wild nodding onion-bulbs)
Camassia spp. Lindl.; Liliaceae (blue camas-bulbs)
Claytonia lanceolata Pursh.; Portulacaceae (spring beauty, mountain potato-corms)
Erythronium grandiflorum Pursh.; Liliaceae (yellow avalanche lily-bulbs)
Fritillaria spp. L.; Liliaceae (rice-root-bulbs)
Lomatium spp. Raf.; Apiaceae (biscuitroots, kous-roots)
Potentilla anserina L.; Rosaceae (silverweed-roots)
Trifolium wormskioldii Lehm.; Fabaceae (springbank clover-rhizomes)

Other plant foods (Total: ~15 species)
Ledum spp. L.; Ericaceae (Labrador-tea-leaves as beverage)
Pinus spp. L.; Pinaceae (pines-inner bark)
Polypodium glycyrrhiza D.C. Eat.; Polypodiaceae (licorice fern-rhizomes as flavouring)
Populus balsamifera L. ssp. trichocarpa T. & G.; Salicaceae (cottonwood-inner bark)
Tricholoma populinum J.E. Lange; Basidiomycete (cottonwood mushroom)

Some aboriginal people make a good living through selling their artworks and other traditional products derived from forest plants. Others supplement their income by these means. Some of these aboriginal products put a different value on plants and plant products that would otherwise be considered "weed species," virtually without value in industrial forestry. Gitxsan Chief Negotiator Don Ryan talks about the "$1,000 Birch Tree."[2] He notes that industrial foresters throughout British Columbia consider paper birch (*Betula papyrifera* Marsh.) to be valueless as a timber species, and have often tried to

TABLE 2. Examples of Traditional Plant Materials of Indigenous Peoples of Northwestern North America (Total, about 135 species) (Turner, 1998)

Wood for construction and manufacture (Total: ~30 species)
Alnus rubra Bong.; Betulaceae (red alder: dishes, masks)
Cornus sericea L.; Cornaceae (red-osier dogwood: sweatlodge frames, basket rims)
Holodiscus discolor (Pursh) Maxim.; Rosaceae (oceanspray: digging sticks, arrows, mat-making needles)
Oplopanax horridus Smith.; Araliaceae (devil's-club: fishing lures)
Taxus brevifolia Nutt.; Taxaceae (Pacific yew: implement handles, digging sticks, wedges, bows, snowshoes)
Thuja plicata Donn.; Cupressaceae (western red-cedar: canoes, house posts, planks, totem poles, boxes, fish weirs)

Wood and other materials for specialized fuel (Total: ~25)
Artemisia tridentata Nutt.; Asteraceae (big sagebrush–bark as tinder)
Betula papyrifera Marsh.; Betulaceae (paper birch–shredded bark as tinder)
Pinus contorta Dougl.; Pinaceae (lodgepole pine–tops as firemaking drills)
Pseudotsuga menziesii (Mirbel) Franco.; Pinaceae (Douglas-fir–bark a hot burning fuel)
Salix lasiandra Benth.; Salicaceae (Pacific willow–wood for firedrill and hearth)
Thuja plicata Donn.; Cupressaceae (red-cedar–wood and shredded bark as tinder, kindling)

Fibrous plant materials (Total: ~25)
Betula papyrifera Marsh.; Betulaceae (paper birch–bark for containers, canoes)
Carex obnupta Bailey.; Cyperaceae (basket sedge–leaves for baskets)
Chamaecyparis nootkatensis D. Don.; Cupressaceae (yellow-cedar–inner bark for baskets, mats, clothing, blankets)
Picea spp. A. Dietr.; Pinaceae (spruce–roots for binding, baskets)
Ribes divaricatum Dougl.; Grossulariaceae (gooseberry–roots for reef nets)
Thuja plicata Donn.; Cupressaceae (red-cedar–inner bark, branches, roots for baskets, mats, clothing, cordage)
Typha latifolia L.; Typhaceae (cattail–leaves for mats, bags, baskets)

Other plant products used in technology (Total: ~40 species)
Abies spp. Mill.; Pinaceae (true firs–boughs for bedding, flooring, incense)
Equisetum spp. L.; Equisetaceae (horsetails–stems as abrasive)
Gaultheria shallon Pursh.; Ericaceae (salal–leafy branches used in cooking pits)
Lysichitum americanum Hultén & St. John.; Araceae (skunk-cabbage–leaves as surface for drying berries, wrapping food, temporary cups)
Philadelphus lewisii Pursh.; Hydrangeaceae (mock-orange–flowers, leaves as cleansing agent)
Pinus spp. L.; Pinaceae (pines–pitch: adhesive, waterproofing)

reduce the numbers of this tree through cutting it down or even spraying it with herbicides. Gitxsan artisans, on the other hand, can make from a single birch tree over $1,000's worth of masks, bowls, spoons, and birch-bark baskets. These are definitely value-added non-timber forest products (see Figure 2).

TABLE 3. Examples of Medicinal Plants of Indigenous Peoples of Northwestern North America (Total, about 200, including those applied in more than one category) (Sources of information: Turner and Bell, 1973; Turner et al., 1981, 1983, 1990; Turner and Hebda, 1991)

General tonics (Total: ~30 species)
Abies spp. Mill.; Pinaceae (true firs–3 spp.–bark)
Alnus spp. Hill.; Betulaceae (alders–3 spp.–bark)
Ledum groenlandicum Oeder.; Ericaceae (Labrador-tea–leaves)
Lonicera involucrata (Rich.) Banks.; Caprifoliaceae (black twinberry, fly honeysuckle–twigs)
Nuphar polysepalum Engelm.; Nymphaeaceae (yellow pond-lily–rhizomes)
Oplopanax horridus (Smith) Miq.; Araliaceae (devil's-club–inner bark, roots)
Shepherdia canadensis L.; Elaeagnaceae (soapberry, russet buffaloberry–twigs)

Purgatives, laxatives, emetics (Total: ~20 species)
Aruncus dioica; Rosaceae (goatsbeard–stems, leaves, roots)
Rhamnus purshiana D.C.; Rhamnaceae (cascara–bark)
Sambucus racemosa L.; Caprifoliaceae (red elderberry–bark, roots)

Salves, poultices and washes for skin ailments (~40 species)
Gaultheria shallon Pursh.; Ericaceae (salal–leaves)
Lysichitum americanum Hulten & St. John.; Araceae (skunk-cabbage–leaves)
Maianthemum dilatatum Wood.; Liliaceae (wild lily-of-the-valley–leaves, roots)
Picea spp. A. Dietr.; Pinaceae (spruces–pitch, bark)
Plantago major L.; Plantaginaceae (broad-leaved plantain–leaves)
Populus balsamifera ssp. *trichocarpa* T. & G.; Salicaceae (black cottonwood)
Rubus spectabilis Pursh.; Rosaceae (salmonberry)

Medicines for colds, coughs, tuberculosis and other respiratory ailments (~40 species)
Achillea millefolium L.; Asteraceae (yarrow–leaves)
Alnus spp. Hill.; Betulaceae (alders–bark)
Artemisia spp. L.; Asteraceae (sagebrush, wormwood–stems, leaves)
Juniperus communis L.; Cupressaceae (common juniper–boughs)
Lomatium nudicaule (Pursh) Colt & Rose.; Apiaceae (wild celery, Indian consumption plant–seeds)
Polypodium glycyrrhiza D.C. Eat.; Polypodiaceae (licorice fern–rhizomes)
Prunus virginiana L.; Rosaceae (choke cherry–bark, fruits)
Pyrus fusca Raf.; Rosaceae (Pacific crabapple–bark)
Tsuga heterophylla (Raf.) Sarg.; Pinaceae (western hemlock–bark)

Aids for internal ailments (digestive tract, internal injuries) (~30 species)
Abies spp. Mill.; Pinaceae (true firs–bark, pitch)
Alnus spp. Hill.; Betulaceae (alders–bark)
Tsuga heterophylla (Raf.) Sarg.; Pinaceae (western hemlock–bark)

Gynaecological medicines (~30 species)
Cornus sericea L.; Cornaceae (red-osier dogwood–twigs, bark)
Equisetum hiemale L.; Equisetaceae (horsetail–stems)
Gaultheria shallon Pursh.; Ericaceae (salal–stems, leaves)
Oenanthe sarmentosa Presl.; Apiaceae (water-parsley–rootstocks)
Sambucus racemosa L.; Caprifoliaceae (red elderberry–bark; TOXIC)

Rheumatism and arthritis (~30 species)
Oplopanax horridus (J.E. Smith) Miq.; Araliaceae (devil's-club–inner bark)
Pinus contorta Dougl.; Pinaceae (lodgepole pine–bark)
Empetrum nigrum L.; Empetraceae (crowberry–twigs)
Taxus brevifolia Nutt.; Taxaceae (Pacific yew–bark)
Urtica dioica L.; Urticaceae (stinging nettle–leafy stems)

Miscellaneous other medicines (~30 species)
Anemone multifida Poir.; Ranunculaceae (Pacific anemone–leaves, stems as poultice; counter-irritant)
Lonicera involucrata (Rich.) Banks.; Caprifoliaceae (black twinberry–twigs, leaves)
Nuphar polysepalum Engelm.; Nymphaeaceae (yellow pond-lily–rhizomes)

NEW DIRECTIONS FOR ABORIGINAL USE OF NTFPS

Many First Peoples are seeking means of employment which would maintain traditional activities and preferred out-door lifestyles and utilize available traditional knowledge while generating a viable income. NTFPs can perhaps provide such an opportunity. There is an increasing awareness of and demand for "natural" forest products, and efforts are beginning to be made to identify issues and opportunities in the development of this industry by communities. For example, in 1998 the Inner Coast Natural Resource Centre in Alert Bay collaborated with the University of Victoria to present a community-sponsored conference on NTFPs which aided the community, including aboriginal participants, in exploring NTFP industry opportunities. A number of aboriginal groups in B.C., such as the Ktunaxa (Kootenai), Secwepemc (Shuswap), Nlaka'pamux (Thompson), and Nisga'a, have on-going research and development programs for NTFPs (Mitchell, 1998). Some studies have also looked at successful attributes of First Peoples' community-based NTFP industries, and provide suggestions for the development of such industries (Chapeskie, 1999; Inner Coast Natural Resource Centre, 1998). Many aboriginal forest-based communities have retained their values, traditions, and practices regarding use of the land, and therefore are in a good position to effectively move into this niche. The resurgence of their own cultures complements well the other opportunities becoming available, especially with activities and programs such as ecotourism, plant propagation and seed production for restoration and landscaping, and activities associated with local museums and cultural centers in which plant use and

TABLE 4. Examples of Aboriginal Art and Food Items, Selections and Prices in Victoria, B.C.* as of Spring, 2000. (Note: RCBM = Royal British Columbia Museum gift shop)

Item	Description	Retail Price
Basket	6″ square, woven cedar bark basket, well constructed but not elaborate, with or without design, sweetgrass finish at top	$110
Basket	3″ round, woven sweetgrass basket with lid	$325
Basket	Very elaborate design with dyed sweetgrass	–2″ version $160 –1″ version $65
Canoe Bailer	Cedar bark with cedar bark ties, enforcers to stiffen, plain (alder?) handle	$55
Canoe Bailer	Same, but–with elaborate handle	$110
Box	Cedar bent-wood box, 8″ with painted design	$1400
Hat	Nuu-Chah-Nulth, sweetgrass, elaborate design	$425
Carvings, Examples	Made with red or yellow cedar, or alder	At RBCM: $100-5000 At Gallery: small $20-50 large $200-8000
Carving	10″ frog and eagle totem of yellow cedar (RBCM)	$2700
Carving	10″ bear of red cedar (RBCM)	$1250
Carving	4″ frog of red cedar (RBCM)	$165
Masks	Full size, made with a combination of: yellow cedar, red cedar, acrylic paint, cedar bark, feather, brass, abalone, alder	At RBCM: $850-6400 At Gallery: up to $5000, average $400-600
Drums	Made with deer hide, yellow or red cedar, feathers, glass; Around 28″ height, 19″ width	$2500-4400
Necklace	"Wildwoman" necklace of wood, feathers, leather and beads	$1825
Necklace	Ceremonial eagle necklace of wood, feathers, cedar bark rope, and beads	$850
Berries	8 oz. jar of soapberries (*Shepherdia canadensis*) (from Wilp Sa Maa'y Wild Berry Co-op, Hazelton, B.C.)	$10.00
Berries	8 oz. jar of huckleberry jam (*Vaccinium membranaceum*) (Wilp Sa Maa'y)	$6.00

* Notes:

The artist usually sets the price of piece, rather than the retailer. Both the Royal BC Museum and the Sa-Nuu-Kwa Gallery claimed that there was higher supply than demand. However they recognized that in order to ensure the continued relationship with good carvers, it was necessary to continue buying on a regular basis. Further, both retail stores found that while products within the thirty to sixty dollar range sold more frequently, more elaborate, expensive items did sell and created a welcome diversity to their selection. Neither retail outlet was willing to supply wholesale costs, however the Sa-Nuu-Kwa Gallery indicated that the usual mark-up was around 100%, but varied according to the price paid to the supplier and periodic sales.

Neither the museum nor the gallery carried any First Peoples' food products. However, the museum claimed that if products which were FDA approved were available, they would be very interested in purchasing them. From previous experience, they have found that products which come in fairly elaborate or "classy" packaging tend to sell best.

FIGURE 2. Secwepemc Elder and basketmaker Dr. Mary Thomas of the Neskonlith Nation, holding a birch-bark basket she has made. She has taught several family members and others in her community the arts of birch-bark and pine needle basketry.

values are featured along with other important cultural knowledge. Using these products in art and filmmaking is also a compatible application.

The benefits of NTFP development reach beyond economic considerations. The increase in use of natural foods in home or restaurant cuisine is not only healthful, but it also increases the awareness of culturally important plants for both the native and non-native consumers (Turner et al. 1995; Craig, 1998). The increased recognition of NTFPs could aid in the lobby to decrease the dependency of communities on timber extraction as the sole economic income from a forest. The increase in awareness of culturally significant plants also supports a viable industry from traditional activities, such as basket-making and berry picking, as demand for natural products grows. Often, too, these products can be marketed together, such as use of special wooden boxes or woven baskets as containers or packaging for wild berry preserves.

For straight resource harvesting, floral greens such as salal (*Gaultheria shallon* Pursh), deer fern (*Blechnum spicant* [L.] Roth.) and sword fern (*Polystichum munitum* [Kaulf.] Presl) can provide an income for individuals in some communities. The salal industry on Vancouver Island, B.C. is estimated to have a value of $20-50 million per year (de Gues, 1995). Most floral greens can be harvested year round, except in the spring when new growth is occurring.

Another very important commercial NTFP is wild mushrooms, particularly pine mushroom (*Tricholoma magnivelare* [Peck] Redhead) and chanterelle (*Cantharellus* spp.). Currently in B.C., around $25-50 million is earned by wild mushroom pickers (some of whom are aboriginal) each year, while $50-80 million is earned by the exporters (Hamilton, 1998). Many mushroom pickers, in order to make a significant amount of money during the mushroom season, follow the maturing of the mushrooms as the season moves from North to South. Andrew Chapeskie, with the Taiga Institute, raises some questions, however, on commercial mushroom picking being a viable economic alternative for aboriginal communities, given these practices: "Would First Nation 'pickers' or 'buyers/distributors' be culturally pre-disposed to become engaged in this practice? What role does territoriality play here and what are its key features in any given setting? Could a possible indigenous preference for adherence to ancestral patterns of territoriality that is expressed in a preference to remain 'local' be addressed by understorey burning?" (Chapeskie, 1999: 6). These are important issues which need to be addressed in the context of each community, and by the people of that community.

Although there are few NTFPs which are marketed at a commercial scale at present, there is an extensive range of species from the forests of western North America which have a potential for use in our society, and the knowledge for their use is still held by many First Peoples. As mentioned, mushrooms and salal are harvested extensively, as are some medicinal plant products such as Pacific yew (*Taxus brevifolia* Nutt.) and cascara bark, St. John's-wort (*Hypericum perforatum* L.; non-indigenous), and bark and roots of Oregon grape (*Berberis* spp.) and devil's club (*Oplopanax horridum* [Smith] Miq.). In B.C. in 1997, it is estimated that the 200-300 commercial gatherers of medicinal plants earned collectively $2-3 million (Canadian $) (Wills and Lipsey, 1999), but most of this would have been from non-aboriginal harvesters. Other forest product uses are also gaining momentum as the popularity of "natural" products increases. For example, companies such as Aveda are promoting teas made with licorice fern rhizome (*Polypodium glycyrrhiza* D.C. Eat.), and the Body Shop specializes in promoting products made with sustainable, alternative forest products. Ecotrust and other environmental organizations have expressed interest in collaborating with aboriginal people in developing such products.

An NTFP with great possibility is the wide array of wild berries which can be found in forests and clearings. In B.C., berries with high use potential include salmonberry (*Rubus spectabilis* Pursh), red huckleberry (*Vaccinium parvifolium* Smith), black huckleberry (*Vaccinium membranaceum* Dougl.), wild blueberries (*Vaccinium* spp.) and salal berries (Turner et al., 1996). Indications are that if the product were available in quantity, there would be a ready market and harvesters would have no difficulty in finding buyers (Turner et al., 1996; Zdenek Tomas, buyer, pers. comm. 1999). However, these berries are small and can take a long while to pick (Juliet Craig found that the time required to harvest 0.5 litres of red huckleberries by two people was 20-28 minutes; Craig, 1998). Also, the return per unit is not high, and the supply somewhat unpredictable. Thus, wild berry-picking may not be an economic alternative for individuals in many cases. Still, value-added processing, such as making jams, jellies, preserves and teas from these berries, provides greater returns for harvesting time.

Purchasers at the Royal BC Museum's giftshop stated that the demand is there for such wild food products, if packaged attractively for tourists (Krista Lousier, RBCM buyer). The Wilp Sa Maa'y Harvesting Cooperative[3] is a wild berry marketing co-operative with an elected Board of Directors, including aboriginal and non-aboriginal members, which has successfully completed its first year of harvesting and production, with two primary products: jarred soapberries (*Shepherdia canadensis* [L.] Nutt.) and black huckleberry jam (see Table 4; see Figure 3). They anticipate expanding their harvest to other species and marketing more widely in the future.

As the demand for "natural products" increases, the market for NTFPs will also increase. However, in order to profit by the current market, and to enhance the increasing trend of alternative forest products use, it would be beneficial for harvesters to educate themselves on the potential. This would include familiarity with processing options and requirements to maximize the value of the product, such as cold storage to maintain the quality of floral greenery. This consideration would allow greater market accessibility. In Mexico City, for example, individuals will pay up to $130 for a wreath made by aboriginal people (Freed, 1998). This type of knowledge also includes methods of utilizing below-grade products harvested (i.e., those which don't meet quality standards set by buyers), as well as convincing companies and individuals of the potentials and worth of the products (Freed, 1998). As noted previously, adding value through processing, will increase the returns to harvesters and decrease the impact of variable seasons and fluctuating prices.

There is also research being done by some aboriginal communities on methods of integrating NTFP harvesting with other activities. For example, the Gitxsan in Old Hazelton are zoning their land base in order to identify and maintain multiple, related land use activities such as logging, NTFP harvesting, and tourism (Gamiet et al., 1998). An Oregon-based basketweaver Pat Courtney Gold, made the important observation, "By sharing our culture [as basketweavers], we help educate others."[4] Authenticity, and cultural traditions embodied in products such as baskets, are highly important elements for tourists and other purchasers of value-added NTFPs.

FIGURE 3. Example of Wilp Sa Maa'y Harvesting Cooperative preserve label.

Wilp Sa Maa'y is "House for Berries" in the Gitxsan language. The Wilp Sa Maa'y Harvesting Coöperative is a community-based initiative to sustainably harvest wild forest products and provide local employment. Wild huckleberries and blueberries are gathered by hand in the mountains of northwestern British Columbia.
To place additional orders, or for more information, please e-mail us at swat@mail.bulkley.net or write us at:
Wilp Sa Maa'y
P.O. Box 354
Hazelton, B.C.
WILP SA MAA'Y
250 ml
Wild Huckleberry Jam
Ingredients: Wild black huckleberries (Vaccinium membranaceum) and wild blueberries (V. Alaskense or V. Ovalifolium), sugar, apple, orange juice, orange peel, lemon peel, cinnamon.
Keep refrigerated after opening
Processed in South Hazelton
British Columbia
Product of Canada

ISSUES AND CONCERNS OF ABORIGINAL PEOPLES IN MARKETING NON-TIMBER FOREST PRODUCTS

Regulating NTFPs, Land Tenure, Access to Resources

At the moment, there is no legislation in British Columbia which deals with NTFPs. There is provision for regulation of NTFPs under the Forest Practices Code, but to date, no regulations have been made at a provincial level. Thus, there is currently no particular restriction on commercial gathering of NTFPs within provincial forest lands and tree farm license areas, whether these are on a First Nation's traditional territory or not. This means that there is no limit to the quantity of a plant product which may be removed, nor in the methods used in harvesting it. Without regulation, it is difficult to ensure that harvesting will be done in an equitable and sustainable manner. Even if studies are made which indicate which specific harvesting techniques should be used in order for sustainability to be maintained, enforcement of these methods will be an ongoing problem. Some provincial Forest Districts, such as Chilliwack and Mission, tried to address such concerns by requiring permits for recreational and commercial botanical forest product harvest. This provided an opportunity to educate and regulate, as well as acquire an inventory of how much plant product was being removed from the woods. The permits, however, were not enabled through statutes and thus could not stand if questioned. Also, the permits were free of charge whereas costs of implementing and monitoring the regulation rose. As both meeting costs and enforcement were difficult, this system was withdrawn (Chilliwack Forest District, 1998).

With a growing awareness and interest in NTFPs, the number of harvesters in the field is increasing, with a resulting competition for land and products. This means that traditionally used territory may be affected by new harvesters. This is a particular problem for aboriginal people, who may have harvested plant and animal resources from a traditional territory for generations, yet now are finding their assets depleted by commercial harvesters from outside the region. An example of such harvesting impacts is with Pacific yew (*Taxus brevifolia*), whose bark is a major source of the anti-cancer drug Taxol. Since Taxol was discovered, and clinical trials initiated by the U.S. National Cancer of Health, the National Cancer Institute, and the pharmaceutical company Bristol-Myers Squibb, yew bark has been harvested in large quantities throughout the Northwest Coast area, and many large yew trees have been killed in the process, some of them poached from parks and Native reserve lands. Pacific yew is highly valued by aboriginal peoples, not only for its medicinal qualities but also for its tough, resilient wood which is used for carving and making of implements such as bows, wedges, harpoon shafts and digging sticks (Turner, 1998). Aboriginal people who use it do so with care and respect. In using the bark for medicine, they apply the traditional technique of harvesting a single strip from the trunk, without girdling it, thus keeping it alive (Turner and Hebda, 1992). Outside harvesters are neither as knowledgeable nor as caring in their removal of the bark, and as a consequence, yew trees, especially large ones, have been virtually eliminated in some regions (Hartzell, 1991).

As with the Pacific yew example, lack of awareness among the new harvesters of sustainable methods of harvesting further increases the problem of outsider exploitation of resources, as it impacts the management of the ecosystem. Also of great concern is the exclusion of aboriginal people from some protected areas which may have traditionally been within their gathering territory: provincial and national parks, wilderness areas and ecological reserves, Department of National Defence lands, and private lands. One example is the case of a Stl'atl'imx (Lillooet) woman who, in August 1997 was charged with picking soapberries (*Shepherdia canadensis*) in Banff National Park, Alberta. The Banff National Park Board decided to waive her fine. However, they set trial dates for August 31 to September 4, 1998 in Alberta. Although she was from B.C. she was required by the courts to hire an Alberta defence lawyer. She had been picking soapberries for 17 years in the Banff area, for food and medicinal purposes. She had been charged for picking berries the previous year, and had been fined $100 and given a court date, but just prior to the court date she was told the Park Board dismissed the charges and would refund the fine. Other aboriginal people have also been charged for picking salal and other plant harvesting activities inside park boundaries.

In some cases, as with BC Hydro rights-of-way, access can be arranged through special agreements. Provisions for co-management of parklands, including several new parks being developed in British Columbia (e.g., the Kitlope Valley Protected Area [Husduwachsdu; between the Province and the Haisla Nation]), and Gwaii Haanas National Park Reserve on Haida Gwaii (Queen Charlotte Islands) (between Canada and the Haida Nations), also include traditional harvesting rights by aboriginal peoples, but may not allow commercial exploitation of the resources of these areas. In the future, it may be possible to work out such arrangements if sustainable harvesting methods are demonstrated and quotas, licenses or other ways to control harvests are developed.

Protecting Intellectual Property Rights and Cultural Integrity

The development of a greater array of NTFPs has many positive potentials, but it also evokes the controversial and increasingly discussed issue of intellectual property rights (IPR). It is crucial to the cultural and economic interests of indigenous peoples, and a matter of respect, that all knowledge sources used in developing and promoting NTFPs be acknowledged, and that people are consulted and appropriately compensated for their knowledge. Aboriginal peoples should be part of NTFP ventures at the outset, and should have an active role in planning and decision-making.

IPR can be a difficult subject as ownership of knowledge is debated between researchers, corporations, and First Peoples. The issues of ownership, however, reach far more deeply than identifying who facilitated the research. The basis of ownership of knowledge in a study that yields economic or other benefits must be linked back to the data on which that study is based, as "ownership is in reality an exercise of a property right (i.e., who is able to exert control over the data)" (Macrina, 1995:193). It is very clear who owns the data in almost all cases–the data is the knowledge of a people, and these people have ethical ownership rights to this knowledge and its use. The knowledge held by an aboriginal person or community is based on centuries of observation, monitoring, experimentation and practical application. The mere recording and transcription of part of this knowledge does not, therefore, bestow ownership to the researcher. However, based on copyright laws, a community actually has no legal ownership of their knowledge, as copyrights at present cannot be granted to a corpus of cultural knowledge, and nor are copyrights in effect until the knowledge is in written form (Greaves, 1994). Of course, traditional knowledge is by definition collective and communal. However, another legal avenue by which First Peoples can retain legal rights to their knowledge is contract law. Under such law, researchers and corporations anticipating benefits from traditional knowledge research sign an agreement with designated community officials, such as Chief and Council of an aboriginal community, in which the terms and conditions surrounding the sharing of knowledge are defined and agreed upon at the outset. Such agreements are relatively common among academic researchers working with aboriginal peoples, but are seldom used by industrial bodies such as pharmaceutical companies.

A "classic" case of cultural appropriation for profit is in the development and marketing of a herbal medicine product called "Original Indian Essence." The package describes the product as: "Genuine nine herb formula of dry native Indian Tea Blend with mild bitter qualities–nature's harmony of herbal riches."[5] The ingredients are: "*Burdock root (*Arctium lappa* L.), *Sheep sorrel herb (*Rumex acetosella* L.), Slippery elm bark (*Ulmus rubra* Muhl.), *Turkish rhubarb root (*Rheum palmatum* L.), *blessed thistle herb (*Carduus benedictus* L.), Mistletoe leaves [*Phoradendron flavescens* (Pursh) Nutt., fam. *Viscum album* [sic]), Kelp (*Laminaria digitata* [sic]), *Watercress (*Nasturtium officinale* R. Br.), *red clover (*Trifolium pratense* L.)." Notably, six out of the nine ingredients, those marked with *, are species introduced to North America, and thus hardly traditional aboriginal remedies as claimed.

On the other hand, there are examples of ethical collaborative ventures in which aboriginal participation is marketed as part of a product. One is the Wabauskang Wildfruits company,[6] which produces wild blueberry products, and advertises: "Wild Blueberry Spreadable Fruit is made to a traditional recipe to capture the unique flavour of wild blueberries hand-picked by Native people in Ontario's north woods." Hopefully, more of the latter type of NTFP businesses will develop, and fewer of the former. Currently, in British Columbia and elsewhere, there are a number of partnerships forming in research and economic development with First Peoples, universities, government and private industry that will lead the way to more, similar ventures in our region. On a related note, there are also examples of native plant nurseries and seed companies being developed and operated by aboriginal entrepreneurs.

Conservation of Forest Species and Ecosystems

The close relationship with the land that the First Peoples held and, in many areas, still hold, allowed for the harvesting of NTFPs to be sustainable. Stewardship was an integral part of the harvesting, and thus rather than depleting the species, growth and propagation was encouraged. The First Peoples of northwestern North America were not simply "gatherers"; they modified and improved the ecosystem around them through a variety of techniques, including selective harvesting, propagation, habitat enhancement, monitoring, pruning, and cultural controls against wastage and irresponsible use. The key to true sustainability was that "plants were not merely 'gathered' from the 'wilderness', but were tended and looked after to promote and enhance populations for the following years" (Craig, 1998:123). This concept is crucial for the current commercial harvesting of NTFPs–the harvesters must be more than simply gatherers, they must tend their forest "gardens."

There is every indication that a NTFP industry could be a sustainable and healthy one for the forest. For example, according to Juliet Craig's interviews with Ahousaht elders (1998), picking berries does not reduce the crop yield for the next year. Further, methods were designed to promote berry production. The tops of the bushes were broken off, which not only allowed for faster picking, but it was also a form of pruning which increased production for the following year. Care must be taken in any large-scale berry harvesting program that the needs of birds, bears and other wildlife are not compromised, and that some areas are left intact for these other users of forest products. Aboriginal people are particularly conscious of such requirements.

Another example of cultivating wild plants is the use of digging sticks for gathering root vegetables, a process which tilled and aerated the soil and enhanced the ground for the roots and other underground parts. Fragments of the "roots" were often left or replanted. Thinning of density-dependent species such as slough sedge (*Carex obnupta* Bailey) for basketry aids in the growth and reproduction of the plants (Craig, 1998). These methods should be studied and encouraged.

Richard Ross (1998) claims that harvesting of salal greens can be sustainable and can be undertaken annually, if the pruning is done correctly. Some sites have been harvested since the 1950s and are still producing good quality stems. However, the level of salal harvesting has increased dramatically over the last five years, thus competition is greater. This has led to higher harvesting requirements and more pressure on the sites and populations. More research is required for this and other NTFP species to determine sustainable levels of harvest, and also, the extent to which the harvesting of salal greens impact the productivity of the berries.

One possibility to help assess harvesting impacts is to develop an Index of Sustainable Harvesting Potential. We are looking at the feasibility of using such an Index to enable a method of predicting what harvesting methods would be optimal for a species and how extensively a population or plant product can be harvested, if at all. It is intended to serve as a guide to be applied to plant harvesting for individual ecosystems and under specific conditions. The preliminary numerical values obtained for the index are based on three scales of specific attributes: 1. habit or growth form, 2. reproductive capacity and life cycle, and 3. frequency and abundance. Other factors would also have to be considered when developing final values for the Index, such as the part of the plant harvested, time of year of harvesting, and the abundance of local populations.

Such an Index could be used in the development of education and legislative guidelines for harvesting NTFPs. In establishing the guidelines, it will also be necessary to consider and prioritize First Peoples' access to the land and species. It is important that harvesters, buyers, and policy makers recognize the need to develop the industry with a solid, well-researched base that will allow it to continue to function ecologically as well as economically. There is a very real possibility that if research, education and regulation are not developed and applied effectively, the NTFP industry will ultimately follow the downward productivity trend of the timber industry.

One of the biggest problems with maintaining stewardship of native forest species, however, is the decreasing habitat due to such human activities as urbanization and industrial development, especially forestry, agriculture and mining. The First Peoples are left with very little of their original land-base and therefore have increasing difficulty in accessing, monitoring, and cultivating traditionally used species. This becomes an even greater issue as NTFP harvesting increases and traditional territory is encroached upon by external commercial harvesters.

We have developed a set of Guiding Principles relating to sustainable harvesting of NTFPs (Appendix 1). Of these, several, under Cultural/Social Factors, relate directly to aboriginal peoples, particularly:

- Indigenous peoples have developed a variety of conservation and sustainable harvesting practices for culturally important species. These practices are often inextricably linked to peoples' worldview and spiritual values.
- Local knowledge is crucial; so is scientific knowledge.
- Intellectual property rights of Indigenous Peoples must be acknowledged and protected.

These principles may seem obvious, but they form a key to appropriate, equitable and sustainable use of NTFPs by aboriginal and non-aboriginal peoples alike.

CONCLUSIONS

Managers and decision-makers in ecosystem-based forest management are increasingly aware of the requirement for and desirability of collaboration with aboriginal peoples in planning for the use of forest ecosystems, whether for harvesting timber, for NTFP gathering, or for other "non-consumptive" uses of the forest. Aboriginal peoples have major interests in the NTFP industry. They have traditionally used a wide range of these products for both domestic and trade purposes, and they continue to have an abiding cultural and economic stake in this industry, focusing on their own traditional lands and resources. First Peoples have often been alienated from their traditional lands and excluded from land-based economic opportunities. Furthermore, their intellectual property rights have not been acknowledged or respected in many cases. Fortunately, this situation is changing both in Canada and the United States, and all of us who wish to promote the sustainable use of NTFPs will be the richer for their continued active participation in this growing venture.

AUTHOR NOTE

The authors would like to thank the following people for their contributions to the development of this article and the ideas it provides: Chief Arthur Adolph, Xaxl'ep First Nation; Dr. Fikret Berkes, University of Manitoba; Iain Davidson-Hunt, Taiga Institute; Dr. Marianne B. Ignace, Secwepemc Education Institute and Simon Fraser University; Chief Ron Ignace, Skeetchestn First Nation and Simon Fraser University; Mary Thomas, Secwepemc Elder, Neskonlith First Nation; Dawn Loewen and Colin Laroque, University of Victoria; Michael Keefer and Pete McCoy, Ktunaxa Nation; Dave Mannix, Snuneymexw First Nation; Ian Tait (stél′mexw siiyá′y), Business Development and Communications, BC Hydro Aboriginal Relations Department, Vancouver; Dr. Darcy Mitchell, Mitchell Consulting Associates; Dr. Alan Thomson, Pacific Forestry Centre, Canadian Forestry Service; Carla Burton and Marilyn Woodcock, Wilp Sa Maa'y Harvesting Cooperative; and Barbara Wilson, Haida Nation; Dr. Richard Atleo; Arthur Robinson, Federal Lands Forester; Dr. Patrick von Aderkas and Dr. Barbara Hawkins, University of Victoria. They also acknowledge, with gratitude, the many aboriginal plant and language specialists, past and contemporary, whose teachings and contributions have kept their traditional botanical knowledge alive.

NOTES

1. These products all combined to yield an estimated market value for Canada in 1997 of over $240,000,000 (Canadian $) (Duchesne and Davidson-Hunt 1998).

2. Lecture to U.B.C. forestry class, Natural Resources Conservation 451, September 1997, Vancouver, B.C.; also, speaker at Helping the Land Heal Conference on Environmental Restoration, Victoria, B.C., November 1998.

3. Wilp Sa Maa'y Harvesting Society, Box 354, Hazelton, B.C. V0J 1Y0; phone contact: Marilyn Woodcock (250) 842-6511, extension 331.

4. Wasco-Tlingit Artist, Washington State Basketry Gathering, Olympia, Washington, October, 1997.

5. Production and Distribution by: IWF-Indian Wisdom Foundation.

6. Wabauskang Indian Reserve, Box 458, Ear Falls, Ontario Canada P0V 1T0.

REFERENCES

Chapeski, A. 1999. Landscape and Livelihood Non-Timber Forest Products in Contemporary First Nations Economies. Paper presented to Nuu-Chah-Nulth Value-Added Workshop. Port Alberni, B.C. March 22-23.

Chilliwack Forest District. 1998. Botanical Forest Products Bulletin. Ministry of Forests. Issue 01. Chilliwack, B.C. Also at: http://www.for.gov.bc.ca/VANCOUVR/district/CHILLIWA/BFPBulletin_1.html Updated September, 1998.

Craig, J. 1998. "Nature was the Provider": Traditional Ecological Knowledge and Inventory of Culturally Significant Plants and Habitats in the Atleo River Watershed, Ahousaht Territory, Clayoquot Sound. Master's Thesis, University of Victoria.

De Geus, P.M.J. 1995. Botanical Forest Products in British Columbia: An Overview. B.C. Ministry of Forests. Integrated Resources Policy Branch. Victoria, B.C.

Deur, D. and N.J. Turner (editors). (In prep.). "Keeping it Living": Indigenous Plant Management on the Northwest Coast, University of Washington Press, Seattle.

Duchesne, L. and I. Davidson-Hunt. 1998. Presentation to the North American Forestry Commission Meetings, Merida, Mexico. Canadian Forest Service, Great Lakes Forestry Centre, Sault Ste Marie, Ontario, and The Taiga Institute for Land, Culture and Economy, Kenora, Ontario.

Freed, J. 1998. Non-Timber Forest Products: An Overview and Outlook. In: Inner Coast Natural Resource Centre. Non-Timber Forest Products Workshop Proceedings. April 3-5, 1998. Alert Bay, British Columbia, Canada.

Gamiet, S., H. Ridenour, and F. Philpot. 1998. An Overview of Pine Mushrooms in the Skeena-Bulkley Region. Northwest Institute for Bioregional Research. Smithers, B.C.

Greaves, T. (Ed.). 1994. Intellectual Property Rights for Indigenous Peoples: A Sourcebook. Society for Applied Anthropology. Oklahoma City, OK.

Hamilton, E. An Overview of the Current Situation of Non-Timber Forest Products in British Columbia. In: Inner Coast Natural Resource Centre. Non-Timber Forest Products Workshop Proceedings. April 3-5, 1998. Alert Bay, British Columbia, Canada.

Hartzell, H. Jr. 1991. The Yew Tree. A Thousand Whispers. Biography of a Species. Hulogosi, Eugene, OR.

Inner Coast Natural Resource Centre. 1998. Non-Timber Forest Products and Services: Where Do We Go from Here? Report of a Strategic Planning Session. July 18, 1998. Alert Bay. B.C.

Kuhnlein, H.V. and N.J. Turner. 1991. Traditional Plant Foods of Canadian Indigenous Peoples. Nutrition, Botany and Use. Volume 8. In: Food and Nutrition in History and Anthropology, edited by Solomon Katz. Gordon & Breach Science Publishers, Philadelphia, Pennsylvania.

Macrina, F.L. 1995. Scientific Integrity: An Introductory Text with Case Studies. ASM Press. Washington, DC.

Mitchell, D. 1998. Non-Timber Forest Products in British Columbia; The Past Meets the Future on the Forest Floor. The Forestry Chronicle. 74(3):359-362.

Peacock, S. and N.J. Turner. 2000. "Just Like a Garden": Traditional Plant Resource Management and Biodiversity Conservation on the British Columbia Plateau. In: Biodiversity and Native America, edited by Paul Minnis and Wayne Elisens, University of Oklahoma Press, Norman, Oklahoma.

Ross, Richard. 1998. The Changing Furure of NTFP's. In: Inner Coast Natural Resource Centre. Non-Timber Forest Products Workshop Proceedings. April 3-5, 1998. Alert Bay, British Columbia, Canada.

Russo, L. and T. Etherington. 1999. Non-Wood News. An Information Bulletin on Non-Wood Forest Products. Wood and Non-Wood Products Utilization Branch (FOPW), FAO Forest Products Division, March, 1999, Vol. 6.

Scientific Panel for Sustainable Forest Practices in Clayoquot Sound. 1995a. First Nations' Perspectives on Forest Practices in Clayoquot Sound, Report 3, Victoria,

B.C. (Report prepared by 18 panel members, with Secretariat). With Appendices V and VI.

Turner, N.J. 1995. Food Plants of Coastal First Peoples. Royal British Columbia Museum Handbook, Victoria, B.C. (Rev. Handbook, orig. published in 1975 by B.C. Provincial Museum), University of British Columbia Press, Vancouver.

Turner, N.J. 1996. "Dans une Hotte." L'importance de la vannerie das l'économie des peuples chasseurs-pêcheurs-cueilleurs du Nord-Ouest de l'Amérique du Nord; (" 'Into a Basket Carried on the Back': Importance of Basketry in Foraging/Hunting/Fishing Economies in Northwestern North America.") Anthropologie et Sociétiés. Special Issue on Contemporary Ecological Anthropology. Theories, Methods and Research Fields. Montréal, Québec, 20 (3): 55-84. (en français).

Turner, N.J. 1997. Food Plants of Interior First Peoples. (Rev. Handbook, orig. published in 1978 by B.C. Provincial Museum.) University of British Columbia Press, Vancouver and Royal British Columbia Museum, Victoria.

Turner, N.J. 1998. Plant Technology of British Columbia First Peoples. (Rev. Handbook, orig. published in 1979 by B.C. Provincial Museum.) University of British Columbia Press, Vancouver and Royal British Columbia Museum, Victoria.

Turner, N.J. and M.A.M. Bell. 1973. The Ethnobotany of the Southern Kwakiutl Indians of British Columbia. Economic Botany 27(3):257-310.

Turner, N.J., R. Bouchard and D.I.D. Kennedy. 1981. Ethnobotany of the Okanagan-Colville Indians of British Columbia and Washington. British Columbia Provincial Museum Occasional Paper No. 21, Victoria, B.C.

Turner, N.J. and R.J. Hebda. 1990. Contemporary Use of Bark for Medicine by Two Salishan Native Elders of Southeast Vancouver Island. Journal of Ethnopharmacology 229 (1990): 59-72.

Turner, N.J. and D.C. Loewen. 1998. The Original "Free Trade": Exchange of Botanical Products and Associated Plant Knowledge in Northwestern North America. Anthropologica XL (1998): 49-70.

Turner, N.J. and S. Peacock. submitted 2001. "The Perennial Paradox": The Traditional Plant Management on the Northwest Coast. In: "Keeping it Living": Indigenous Plant Management on the Northwest Coast, edited by Douglas Deur and Nancy J. Turner, University of Washington Press, Seattle.

Turner, N.J., S. Philip and R.D. Turner. 1995. Traditional Native Plant Foods in Contemporary Cuisine in British Columbia. pp. 23-30, In: Jo Marie Powers and Anita Stewart (ed.). Northern Bounty. A Celebration of Canadian Cuisine. Random House of Canada, Toronto.

Turner, N.J., R. A. Reed, J. Macko and S. Philip. 1996. Wild Berry Products: Marketing Potential in Southwestern British Columbia. Forest Renewal BC Research Program, Summary Report, Year 1. Victoria, B.C.

Turner, N.J. R.Y. Smith, J.T. Jones, and R.A. Reed. Submitted 2001. "A Fine Line Between Two Nations": Ownership Patterns for Plant Resources Among Northwest Coast Indigenous Peoples–Implications for Plant Conservation and Management. In: "Keeping it Living": Indigenous Plant Management on the Northwest Coast, edited by Deur, D. and N.J. Turner, University of Washington Press, Seattle, in prep., 1999.

Turner, N.J., J. Thomas, B.F. Carlson and R.T. Ogilvie. 1983. Ethnobotany of the

Nitinaht Indians of Vancouver Island. British Columbia Provincial Museum Occasional Paper No. 24

Turner, N.J., L.C. Thompson, M.T. Thompson and A.Z. York. 1990. Thompson Ethnobotany. Knowledge and Usage of Plants by the Thompson Indians of British Columbia. Royal British Columbia Museum, Memoir No. 3, Victoria, British Columbia.

Wills, R., and R. Lipsey. 1999. An Economic Strategy to Develop Non-Timber Forest Products and Services in British Columbia. Forest Renewal BC Project No. PA 97538. Victoria, B.C.

APPENDIX 1

PRINCIPLES OF SUSTAINABLE HARVESTING OF NON-TIMBER FOREST PRODUCTS

A. ECOLOGICAL/BIOLOGICAL FACTORS

- Ecosystem integrity has primary importance.
- Species interact with, and depend upon, each other.
- Species respond differentially to harvesting, depending on a multiplicity of biological and ecological factors.
- Reproductive and regenerative capacity and rate determine/influence sustainable harvesting potential.
- Some species have extremely high ecosystem values (i.e., keystone species); these must be extremely carefully monitored and protected.
- Ecosystems undergo successional changes following disturbance, including large scale disturbance such as burning and logging.
- It is important to recognize genetic (population) diversity and diversity of ecological structure and function, as well as species diversity.
- Maintenance of population characteristics is a fundamental objective (e.g., need to maintain a balance of age classes, maintain the range of genetic variability, and maintain habitats [the biggest threat to biodiversity is habitat loss]).
- Need to consider cumulative effects of harvesting (e.g., combined impacts of harvesting with others such as logging, overgrazing, wetlands depletion, pests, urbanization) in determining carrying capacity of an ecosystem.
- Small, dispersed populations are generally more vulnerable than widespread, large populations (but these can also be at risk, e.g., bison, passenger pigeon).
- Species with low reproductive capacity, low dispersability, and low adaptive capacity are at higher risk from harvesting activities.
- Natural, long-standing ecosystems should be protected against invasive species (weeds or invasive animals); this must be considered in terms of harvesting-related disturbance.

B. HARVESTING

- Harvesting intensity, seasonality and periodicity affect species responses.
- Extensive time periods must be considered in measuring responses to harvesting (i.e., develop harvesting and marketing in a long term perspective, with future generations' needs and opportunities in mind).
- Constant monitoring and adaptive management are essential. Keep careful records, maps, documentation.
- Diversification of products reduces impacts on species and populations.
- Adding value to non-timber forest products is a key to sustainability.
- Harvesting whole, entire plants from the wild is not desirable.
- Harvesting methods should minimize disturbance to natural ecosystems.
- Non-consumptive "use" of products should be encouraged (e.g., photography, ecotourism, educational programs).

C. CULTURAL/SOCIAL FACTORS

- Indigenous peoples have developed a variety of conservation and sustainable harvesting practices for culturally important species. These practices are often inextricably linked to peoples' worldview and spiritual values.
- Local knowledge is crucial; so is scientific knowledge.
- Sustainable harvesting potential should determine marketable product calculations, i.e., what the ecosystem can support, not what the market requires.
- Education, collaboration and agreement on principles of sustainability and mutually agreed upon, mutually applied controls/rules are all important factors. Harvesting should be coordinated to reduce risks of cumulative harvesting impacts.
- Intellectual property rights of Indigenous Peoples must be acknowledged and protected. So must private land ownership.
- Safety of harvesters, and users of non-timber forest products is of paramount importance.

D. MARKETING/ECONOMIC FACTORS

- Need to consider all values (ecological, cultural), not just monetary values, of non-timber forest products. Monetary values should be subservient to ecological and cultural values. Maintain holistic, interdisciplinary approaches to product selection, harvesting and marketing.
- Accessibility is a factor in harvesting.
- Marketing strategies should include consideration of local products and cultural associations.
- Efficient marketing means reducing wastage, proper storage and preservation, local processing and local marketing.
- "Clusters" of compatible products (e.g., health and cleansing products) from one region will improve marketing efficiency.
- Product packaging is of primary importance.

An Overview of Non-Timber Forest Products in the United States Today

Susan J. Alexander
Rebecca J. McLain

SUMMARY. As people become more interested in personal health and family activities, demand for wild forest products has increased. This increased demand coupled with an increased concern for sustainable management practices has focused attention on the variety of issues and products involved in the non-timber forest products industry. Forest management organizations have gradually increased funding for research and management of non-timber forest products over the past two decades. The broad categories of U.S. non-timber forest products include floral greens, Christmas greens, ornamentals and craft materials, wild edibles, medicinals, ceremonials/culturals, and native transplants. The increase in resource pressure has had many policy reactions, including restricted access, harvesting fees, and harvest limits. Opportunities for public input to policy decisions on federal, state and private land are often unclear or nonexistent. Researchers, managers, and policy makers are working to understand the multitude of issues surrounding non-timber forest products, including biology, management, public policy and equity issues. *[Article copies available for a fee from The Haworth Document Delivery Service: 1-800-342-9678. E-mail address: <getinfo@haworthpressinc.com> Website: <http://www.HaworthPress.com>]*

Susan J. Alexander is Research Economist with the USDA Forest Service Pacific Northwest Forest Sciences Laboratory, Corvallis, OR 97331 (E-mail: salexander@fs.fed.us).

Rebecca J. McLain is Co-Founder and Director of the Institute for Culture and Ecology, P.O. Box 6688, Portland, OR 97228 (E-mail: mclain@ifcae.org).

[Haworth co-indexing entry note]: "An Overview of Non-Timber Forest Products in the United States Today." Alexander, Susan J., and Rebecca J. McLain. Co-published simultaneously in *Journal of Sustainable Forestry* (Food Products Press, an imprint of The Haworth Press, Inc.) Vol. 13, No. 3/4, 2001, pp. 59-66; and: *Non-Timber Forest Products: Medicinal Herbs, Fungi, Edible Fruits and Nuts, and Other Natural Products from the Forest* (ed: Marla R. Emery, and Rebecca J. McLain) Food Products Press, an imprint of The Haworth Press, Inc., 2001, pp. 59-66. Single or multiple copies of this article are available for a fee from The Haworth Document Delivery Service [1-800-342-9678, 9:00 a.m. - 5:00 p.m. (EST). E-mail address: getinfo@haworthpressinc.com].

KEYWORDS. Medicinals, floral greens, wild edibles, non-timber forest products

GROWING PUBLIC RECOGNITION OF THE VALUES OF NON-TIMBER FOREST PRODUCTS

Although non-timber forest products (NTFPs) always have been an important source of livelihood in some areas and among some populations in the United States, interest in NTFPs among scientists and forest managers has been low through most of the 20th century. Until the late 1980s, as elsewhere in the world, most U.S. foresters referred to NTFPs as "minor" forest products. The low status of NTFPs among forest management organizations and mainstream scientific communities was reflected in the low levels of funding accorded to the management of these products within public and private forest management organizations and the small number of scientific articles published on NTFPs in the United States prior to 1990. Over the past two decades, however, forest management organizations have gradually increased funding levels for NTFP management and research, and a small but growing body of scientific literature on NTFPs has emerged in mainstream scientific journals. This shift in behavior on the part of forest managers, policy makers, and scientists stems in part from the increased economic demand for NTFPs locally and globally and in part from new scientific understandings about the critical role of species and structural diversity for sustainable forest management.

MAJOR CATEGORIES OF NTFPS IN THE UNITED STATES

The broad categories of nontimber forest products produced in the United States are floral greens, Christmas greens, ornamentals and craft materials, wild edibles, medicinals, ceremonial/culturals, and native plant transplants. As discussed by Freed (this issue), the sale of many of these products has long contributed to local economies in the United States. Although some wild edibles have been commercially harvested for centuries (e.g., maple syrup and berries), others, such as wild mushrooms, have only emerged as important commercial commodities during the past two decades. A brief overview of recent economic trends in the medicinals, floral greens, and wild edible sectors is provided below.

Medicinals

Many traditional medicines disappeared from common usage in the United States following the rise of scientific medicine and the introduction of manufactured pharmaceuticals in the early 20th century (Vance 1995); however, a gradual resurgence of interest in holistic medicine has prompted an increased demand for and interest in wild plants for medicinal purposes. Pacific yew (*Taxus brevifolia* Nutt.) and American ginseng (*Panax quinquefolius* L.) are examples of medicinal plants for which demand has risen sharply in the past decade. The economic value of these products can be substantial. For example, ginseng exports in 1994, most of which were shipped to Hong Kong, were valued at over $75 million (Viana et al. 1996). Prices for ginseng root produced in the United States in 1994 varied from $24 per pound for domesticated ginseng root to as high as $300 for semi-wild ginseng (Viana et al. 1996). Other important medicinal plants harvested in the United States include purple foxglove (*Digitalis purpurea* L.) and maidenhair fern (*Adiantum* spp.) (Savage 1995; Foster 1995). NTFPs gained sustained national attention in the United States as a result of debates over whether and how to allocate access to yew bark on federally managed forests in the Pacific Northwest during the late 1980s and early 1990s. The diversity of species harvested and the lack of knowledge about medicinal plants among many forest land managers complicates policy making and law enforcement for these products.

Floral Greens

Floral and Christmas greens and Christmas ornamentals are an important sector of the nontimber forest products industry. Significant plants in the floral greens industry include salal (*Gaultheria* shallon Pursh), evergreen huckleberry (*Vaccinium ovatum* Pursh), Cascade Oregon-grape (*Berberis nervosa* Pursh), western swordfern (*Polystichum minitum* [Kaulf.] Presl), beargrass (*Xerophyllum tenax* [Pursh] Nutt.), and many moss species. Most species used in the floral greens market are harvested year-round except in spring when the new growth is tender, and in winter when snow makes harvest difficult (von Hagen et al. 1996). Christmas greens are harvested primarily in fall and winter. Commercial species include such trees as noble fir (*Abies procera* Rehder), Douglas-fir (*Pseudotsuga menziesii* (Mirbel) Franco), and western redcedar (*Thuja plicata* Donn.), among others. Other products, such as cones, are harvested for the Christmas ornamental market (Schlosser et al., n.d.; Thomas and Schumann 1993). Many floral greens are harvested for export markets. For example, 80 percent of the floral greens harvested in Washington state are shipped abroad, primarily to wholesalers in Germany and the Netherlands (Savage 1995). A survey conducted of floral and Christmas green producers west of the Cascade mountains in Washington, Oregon and southwestern British Columbia for the 1989 business year identified approximately 60 floral greens businesses, most of which were located in the state of Washington (Schlosser et al. 1991). These businesses employed about 10,300 people, and the value of floral greens sold at the point-of-first-sale was estimated at US $128.5 million for 1989 (Schlosser et al. 1991).

Wild Edibles

Markets for many wild edibles, including wild berries and fruits, nuts, tree sap and fungi have expanded over the past two decades. Wild blueberries (*Vaccinium angustifolium*) and big huckleberry (*Vaccinium membranaceum* Dougl.) are two of the more popular commercially harvested species of wild berries (Thomas and Schumann 1993). Blueberry and huckleberry picking are also important recreational activities in both the Midwest and eastern United States (blueberries) and the Pacific Northwest (huckleberries). A few national forests in the Northeast, Midwest, and Pacific Northwest have recently initiated berry management regimes involving understory burning and thinning of timber as a means to enhance berry production in areas that were managed as berry fields by Native Americans prior to the Euro-American occupation (Thomas and Schumann 1993; Alexander et al. this issue).

Maple syrup production, which involves refining the sap of the sugar maple (*Acer saccharum* Marsh.), has been an important economic activity in the Northeastern and Midwestern states for centuries. Approximately 4.1 million liters of maple syrup were produced in the United States in 1995, and the value of this production was estimated at $25 million (Viana et al. 1996). Forests in areas with high populations of sugar maples are thinned to remove undesirable species and to induce maple crown formations that favor greater sap production (Viana et al. 1996).

Since the mid-1980s, the wild mushroom industry has gained increasing importance as an NTFP industry in the United States. The expansion of the wild mushroom industry in the United States is closely linked to declines in wild mushroom habitat and productivity in areas that formerly had supplied European and Japanese markets (de Geuss 1992; Denison and Donoghue 1988; Molina et al. 1993). Although the vast majority of wild mushrooms harvested in the United States are exported to either Europe or Japan, demand within the United States is also rising. A combination of factors has led to the rise in domestic interest in wild mushrooms: (1) increasing interest in gourmet food among United States consumers; (2) a decrease in consumer fears about the dangers of mushrooms in general as the use of exotic cultivated mushrooms expands; and (3) increased opportunities for well-capitalized foreign companies to expand into the United States' wild mushroom markets following reductions in trade barriers with the implementation of the North American Free Trade Agreement (NAFTA) and General Agreement on Tariffs and Trade (GATT). The entry of foreign capital has stimulated tremendous growth in the United States' wild mushroom industry; however, the impacts of this growth on U.S.-based companies have not been documented. The bulk of commercial wild mushrooms are probably harvested in the Pacific Northwest. Commercial harvesting of certain species also occurs in the Midwest, the South, and the Rocky Mountain West. The four species most commonly harvested commercially in the Pacific Northwest include chanterelles (*Cantherellus* spp.), morels (*Morchella* spp.), boletes (*Boletus* species), and matsutake (*Tricholoma magnivalare* [Peck] Redhead) (Schlosser and Blatner 1995). The gross value for production of wild mushrooms in the three states of Washington, Oregon, and Idaho in 1992 was US $41.1 million (Schlosser and Blatner 1995).

EMERGING ISSUES IN NTFP MANAGEMENT

The expansion of NTFP industries in response to demands for new products and larger quantities of previously harvested products has generated public debate on how NTFPs should be managed. The entry of more and new types of harvesters and the subsequent growth in demand for access to NTFPs has been accompanied by demands for policies to restrict access, impose or raise harvesting fees, prohibit unsustainable harvesting practices, and limit the quantities and types of NTFPs removed from public and private forests. Some regional plant resources are in limited supply or threatened by habitat disturbance (Hartzel 1991; Molina et al. 1993), raising concerns about the possible negative ecological consequences of unregulated NTFP harvesting.

Scientific evidence about whether current rates of NTFP harvesting are unsustainable is, however, non-existent or limited for most species and products. Scientific information is lacking about many critical aspects of NTFP species, including estimates of abundance, habitat requirements, growth and yield characteristics, susceptibility to insects and disease, and responses to harvesting (Minore 1994; Moore 1993). And while harvesters often have considerable knowledge of these factors for the species they harvest, it is increasingly apparent that current public policy making processes are structured in ways that do not encourage non-timber forest product harvesters and buyers to participate in the development of policy decisions and regulations.

NTFP KNOWLEDGE PRODUCTION AND STAKEHOLDER INVOLVEMENT

As debates over NTFPs gain visibility in the United States, support and interest for producing and incorporating knowledge about NTFPs into forest management decisions has grown. The following papers in this issue describe examples of the two very different, but potentially intersecting, approaches that are being used to develop and utilize knowledge about NTFPs in the United States. These consist of: (1) Development and incorporation of scientific knowledge into management decisions and (2) Recognition of the value of harvesters' knowledge and the importance of involving harvesters in forest management decisions.

During the past decade, a variety of scientific research projects have been initiated to help answer some of the questions that have been raised about the long-term sustainability of NTFP harvesting and to identify mechanisms for expanding harvester and buyer involvement in NTFP policy decisions. Much of this research has taken place in the Pacific Northwest, a region where media attention to NTFPs has been high and advocates of greater regulation of NTFPs have been very vocal. The USDA-Forest Service's Pacific Northwest Research Station, which has several branches in the states of Washington and Oregon, has been one of the strongest supporters of NTFP research since 1989. Section two of this issue thus highlights projects that scientists at the Pacific Northwest Research Station have conducted or are proposing to carry out in order to provide more information about the biological, social, and economic aspects of NTFPs in the Pacific Northwest Region.

Though scientific knowledge about NTFPs is clearly an important element for developing sustainable NTFP management policies in a post-industrial society such as the United States, the knowledge and political participation of those who engage in NTFP harvesting, buying, and processing for a living is equally critical. Section three thus shifts attention toward the harvesters and buyers of NTFPs–their knowledge, stewardship values, social interactions, tenure conflicts, and efforts to influence forest policy.

REFERENCES

Alexander, S.J., R.J. McLain and K.A. Blatner. 2001. Socio-economic research on non-timber forest products in the Pacific Northwest. J. Sust. For. 13(3/4): 95-103.

de Geus, N. 1992. Wild mushroom harvesting session minutes. In: proceedings of a conference on wild mushroom harvesting. Ministry of Forests, Integrated Resources Branch (March 3), Victoria, BC.

Denison, W.C. and J. Donoghue. 1988. The wild mushroom harvest in the Pacific Northwest: past, present, and future. Unpublished manuscript.

Foster, S. 1995. Forest pharmacy: medicinal plants in American forests. Forest History Society, Durhan, NC.

Hartzel, H. Jr. 1991. The yew tree–a thousand whisper. Hulogosi, Eugene, OR.

Minore, D. 1994. Needed basic biological research on commercially important plants. In C. Schnepf (ed.). dancing with an elephant: proceedings of the business and science of special forest products. Western Forestry and Conservation Association, Portland, OR.

Molina, R., T. O'Dell, D. Luoma, M. Amaranthus, M. Castellano and K. Russell. 1993. Biology, ecology, and social aspects of wild mushrooms in the forests of the Pacific Northwest: a preface to managing commercial harvest. Gen. Tech. Rep. PNW-GTR-309. U. S. Department of Agriculture, Forest Service, Pacific Northwest Research Station, Portland, OR.

Moore, M. 1993. Medicinal plants of the Pacific West. Red Crane Books, Santa Fe, NM.

Savage, M. 1995. Pacific Northwest special forest products: an industry in transition. Journal of Forestry 93(3):6-11.

Schlosser, W., K.A. Blatner and D.M. Baumgartner. [n.d.] Floral greens and Christmas ornamentals: important special forest products. Washington State University Cooperative Extension, Pullman, WA.

Schlosser, W., K.A. Blatner and R.C. Chapman. 1991. Economic and marketing implications of special forest products harvest in the coastal Pacific Northwest. Western Journal of Applied Forestry 6(3):67-72.

Schlosser, W. and K.A. Blatner. 1995. The wild edible mushroom industry of Washington, Oregon, and Idaho: a 1992 survey of processors. Journal of Forestry 93(3):31-36.

Thomas, M.G. and D.R. Schumann. 1993. Income opportunities in special forest products: self-help suggestions for rural entrepreneurs. Agricultural Information Bulletin 666. U. S. Department of Agriculture, Forest Service, Washington, DC.

Vance, N.C. 1995. Medicinal plants rediscovered. Journal of Forestry 93(3):8-9.

Viana, V.M., A.R. Pierce and R.Z. Donovan. 1996. Certification of non-timber forest products. In V. Viana, J. Ervin, R. Donovan, C. Elliott and H. Gholz (eds.). Certification of forest products: issues and perspectives. Island Press, Covelo, CA.

von Hagen, B., J.F. Weigand, R. McLain, R. Fight and H. Christensen (comps.). 1996. Conservation and development of non-timber forest products in the Pacific Northwest: an annotated bibliography. Gen. Tech. Rep. PNW-GTR-375. U. S. Department of Agriculture, Forest Service, Pacific Northwest Research Station, Portland, OR.

Non-Timber Forest Products in Local Economies: The Case of Mason County, Washington

James Freed

SUMMARY. Non-timber forest products (NTFPs) have been a vital part of the local economies of Mason County, Washington since the first peoples came there over 9,000 years ago. First Americans used NTFPs in every fact of their lives. The new Americans, from early Euroamericans to the newest Asian Americans, have used nontimber forest products to provide subsistence resources and income support. Beginning in the 1970s, increased demand for medicinals, wild mushrooms, and floral products brought Mason County's NTFP industries back into the limelight. Unfortunately, the rise in demand for NTFPs has increased social conflict in Mason County. Indeed, disputes over harvesting practices and the tension between floral greens and wild mushroom business over across to NTFP leaves have made Mason County the floral point of recent efforts to expand government regulation of the NTFP industry in Washington. However, NTFPs may also provide opportunities for decreasing the political conflict over timber management in the region by creating financial incentives for landowners to maintain longer timber rotations. *[Article copies available for a fee from The Haworth Document Delivery Service: 1-800-342-9678. E-mail address: <getinfo@haworthpressinc.com> Website: <http://www.HaworthPress.com> © 2001 by The Haworth Press, Inc. All rights reserved.]*

James Freed is Special Forest Products Extension Professor, Washington State University, P.O. Box 47037, Olympia, WA 98504-7037 (E-mail: freedj@wsu.edu).

[Haworth co-indexing entry note]: "Non-Timber Forest Products in Local Economies: The Case of Mason County, Washington." Freed, James. Co-published simultaneously in *Journal of Sustainable Forestry* (Food Products Press, an imprint of The Haworth Press, Inc.) Vol. 13, No. 3/4, 2001, pp. 67-69; and: *Non-Timber Forest Products: Medicinal Herbs, Fungi, Edible Fruits and Nuts, and Other Natural Products from the Forest* (ed: Marla R. Emery, and Rebecca J. McLain) Food Products Press, an imprint of The Haworth Press, Inc., 2001, pp. 67-69. Single or multiple copies of this article are available for a fee from The Haworth Document Delivery Service [1-800-342-9678, 9:00 a.m. - 5:00 p.m. (EST). E-mail address: getinfo@haworthpressinc.com].

KEYWORDS. Native Americans, medicinals, mushrooms, floral greens, wild edibles, forest management

NTFPs have long been integral to the cultures and economies of the peoples living in the mountain and lowland forests of Mason County, Washington. Situated on the west side of Puget Sound, Mason County is home to one of the world's most productive temperate rain forests, and is known around the nation for the quality of its timber. Though less widely known, the county is also home to some of the nation's most prosperous NTFP businesses, particularly in the floral greens sector.

Just how long have NTFPs been important to the inhabitants of Mason County? Very likely since the first humans settled this beautiful land. Native American peoples of the Mason County area relied heavily on NTFPs, which they gathered for subsistence use and for trade with other peoples. These plant materials were used in the making of containers (baskets, boxes), clothing (skirts, shirts, dresses), cooking utensils (bowls, spoons, forks), medicinals (high blood pressure, diabetes, stress, constipation), coloring (dyes, paints, stains), edibles (fruits, nuts, roots, vegetables), and beauty care (soap berry, nettles, lichens).

The Euroamericans who occupied the region in the 1800s used a wide variety of NTFPs for many of the same purposes as the Native Americans. As the Mason County economy became increasingly integrated into the industrial economies of the eastern United States and Europe, manufactured goods and medicines began to replace those gathered from the forest. The development of a thriving floral greens business in Mason County in the early 1900s, however, allowed NTFPs to remain an important source of income for logging and fishing families and for residents who could not or did not participate in more conventional occupations. During the 1930s depression, many Mason County residents survived by participating in the harvesting, processing, and trade of floral greens, wild berries, and wild mushrooms.

Beginning in the 1970s, increased demand for medicinals, wild mushrooms, and floral products brought Mason County's NTFP industries back into the limelight. Wild vegetables such as fiddlehead fern, camas roots, and water cress harvested in the county's forests now appear in regional markets. National and international demand for a wide variety of the area's lichens, mosses and liverworts for medicinal purposes is growing. Local and international demand has also increased for wild mushrooms and berries, both of which are found in abundance in Mason County. The combination of a wet, mild climate and an abundant supply of 25-80-year-old Douglas fir stands make Mason County one of the nation's most productive sites for wild chanterelles, and a variety of lesser known wild edible fungi.

The rise in demand for NTFPs has increased social conflict in Mason County. During the past ten years, hardly a fall season goes by without a newspaper article or two about cases of NTFP poaching and tensions over NTFP access rights in the area's forests. Indeed, disputes over harvesting practices and the tension between floral greens and wild mushroom businesses over access to NTFP leases have made Mason County the focal point of recent efforts to expand government regulation of the NTFP industry throughout Washington State. At the same time, however, NTFPs offer the possibility for decreasing the intense political conflict over timber management that has plagued the area since the 1980s. Joint production of timber and NTFPs, for example, could provide the financial incentives for landowners to maintain longer timber rotations while encouraging biological and structural diversity in the area's forests and providing a range of economic opportunities for NTFP harvesters.

SECTION II: RESEARCH ON NON-TIMBER FOREST PRODUCTS IN THE PACIFIC NORTHWEST

Research in Non-Timber Forest Products: Contributions of the USDA Forest Service, Pacific Northwest Research Station

Nan C. Vance

SUMMARY. Non-timber forest products (NTFPs) have emerged as a complex set of issues reflecting changes in society and how natural resources are regarded. These issues range from the sustainability of forest management practices to the relationship of diverse cultures and communities to public lands and their resources. Research and its relationship to this set of issues is a relatively unknown aspect of NTFPs.

Nan C. Vance is Team Leader, Biology and Culture of Forest Plants Team, USDA Forest Service, Pacific Northwest Research Station, Forestry Sciences Laboratory, 3200 SW Jefferson Way, Corvallis, OR 97331.

[Haworth co-indexing entry note]: "Research in Non-Timber Forest Products: Contributions of the USDA Forest Service, Pacific Northwest Research Station." Vance, Nan C. Co-published simultaneously in *Journal of Sustainable Forestry* (Food Products Press, an imprint of The Haworth Press, Inc.) Vol. 13, No. 3/4, 2001, pp. 71-82; and: *Non-Timber Forest Products: Medicinal Herbs, Fungi, Edible Fruits and Nuts, and Other Natural Products from the Forest* (ed: Marla R. Emery, and Rebecca J. McLain) Food Products Press, an imprint of The Haworth Press, Inc., 2001, pp. 71-82. Single or multiple copies of this article are available for a fee from The Haworth Document Delivery Service [1-800-342-9678, 9:00 a.m. - 5:00 p.m. (EST). E-mail address: getinfo@haworthpressinc.com].

This paper reports on early NTFP research by scientists in the USDA Forest Service's Pacific Northwest Research Station. It characterizes efforts over approximately five years and identifies their key elements. It also discusses the role research has and could play in addressing the problems and questions associated with NTFPs and sustainable forestry.

KEYWORDS. Pacific Northwest, non-timber forest products, USDA Forest Service, native forest plants, native forest fungi

INTRODUCTION

Non-timber forest products (NTFP) have emerged as a complex set of issues with innumerable implications for sustainable forestry. These issues reflect changes in how society regards natural resources. They range from managing natural resources for multiple values to questions of sustainability under ecosystem-based management and the relationship of diverse cultures and communities to public lands and their resources. This presents a clear challenge for the research enterprise to renew itself if it is to meet the changing needs and expectations of the society that supports its efforts (Committee for Economic Development 1998).

Sustainable management of forests, whether under federal or private land ownership, depends on a flow of updated information from easily accessible scientific resources (National Research Council 1998). A constant challenge for researchers in addressing a new set of issues is that they compete for attention with existing problems already on the research agenda. Thus, it is useful to inquire how a research enterprise, such as that found in the research branch of the Forest Service, begins to address new issues and bring new knowledge to bear on them. One way this may occur is through flexibility in redirecting or expanding research priorities, which allows scientists to respond to emerging issues. The existing research framework might also be sufficiently broad and robust to accommodate new inquiry along non-traditional lines, or scientists may simply risk asking new but potentially rewarding research questions. New issues and problems provide opportunities for scientists to be at the cutting edge in advancing a state of knowledge. The degree that scientists within an institution can create new knowledge may indicate the degree to which an institution supports an environment that fosters independence and creativity among its scientists (Committee for Economic Development 1998).

In this regard, NTFP research poses an intriguing paradox. The issue is so new that it provides great opportunity for generating new knowledge. However, it does not have a sound body of knowledge on which to build and the relevancy of information could be short lived as needs for information rapidly evolve. NTFPs are part of a larger natural resource arena where research appears to be increasingly conducted in a politically volatile environment. The drive for rapid production of research results to be incorporated into policy challenges researchers to maintain the integrity of their distinct roles. Indeed, the fundamental relationship of research to natural resources policy within the agency is being questioned as the boundaries dividing the separate roles of research and policy have become less distinct (Mills et al. 1998).

In the Pacific Northwest, NTFPs have been a consistent if not significant component of public lands resource use. Trade in forest products that furnish decoratives and greens has provided a principal and supplementary income for several generations of harvesters. The United States Department of Agriculture's (USDA) Forest Service published the first study of this trade in the Pacific Northwest at least 50 years ago (Shaw 1949). A renewed and expanding interest in NTFPs in the past decade has engaged Forest Service scientists sufficiently to afford the opportunity to examine and characterize this research activity. Drawing on results from a survey of scientists in the Pacific Northwest Research Station, this paper seeks to identify key elements of research that addressed NTFPs from 1992 to 1997 in the Forest Service's largest research unit.

NTFP RESEARCH-CONTEXT AND APPROACH

General understanding of NTFPs and scientific recognition are nascent. One characteristic of research on such an emerging issue is a general lack of knowledge at so many levels that the research framework and history available to build upon is scarce or lacking. Because of this, new information will tend to have a greater impact than that of a well-researched subject. The perception of being on the "cutting edge" of an issue also generates new research interest. Ideally, the exploratory nature of this research should create an environment of inclusion that will expand the breadth of information from sources usually considered irrelevant to forestry. In this research environment, integration and synthesis can become sophisticated and important tools.

The conservation and sustainable management of hundreds of commercially important plant and fungal species will depend upon strategies for sustainable harvest and ecosystem protection. The challenge for traditional forestry research, is to participate in broadening the notion of forests so that the understory and forest floor (where most NTFP are found) will be fully considered in the discussion of forestry values. Under such an approach, NTFP plant and fungal species would be studied at multiple levels to evaluate their functional role in habitats and ecosystems. New information on organisms would support development of more rigorous conservation and monitoring protocols and provide new insights into ecosystem components and the effects of human disturbance.

In addition, new and creative approaches for managing diverse resources on public lands may use strategies designed to involve "stakeholder" communities. Using international models of conservation and product-led community development, consideration would be given to co-management schemes to help managers monitor resources, reduce illegal harvest, address multi-cultural conflicts and achieve equitable access to forest resources (Jungwirth and Brown 1997). The grass-roots nature of nontimber forest resources offers novel opportunities for creating new models for using diverse resources while meeting stewardship responsibilities to the nation's forests (Molina et al. 1997).

This suggests a need for greater inclusion of disciplines such as ethnobotany, sociology, economics, anthropology, and botany to complement and integrate with other forestry-related disciplines (Vance 1997). Aided by development of new information technology, research should create a more thorough understanding of how society and forests interact. With this expanded research approach, biodiversity and multiple resource use will have a broader conceptual basis more in line with the fuller range of forest uses being recognized internationally (Clay 1996, Freese 1997).

Research may address NTFPs from two primary perspectives, both valid, but having different outcomes. The first perspective places the species from which a product is derived in an ecological context. In that context harvesting species may raise questions that include concerns about resource sustainability, ecological relationships, specific habitats, effects on population viability and reproduction. Analysis of resource and ecosystem sustainability or the potential risks of commercial exploitation tends to produce information that would result in the following:

- Increased ability to assess commercially and ecologically important species for harvest impacts on the organism, community, and related ecosystem viability.
- Improved native plant conservation strategies and analyses for land managers and conservationists.
- Improved management of biodiversity based on integrated knowledge of social, economic, and terrestrial ecosystem processes.
- Use of native NTFP species in habitat restoration.

With respect to understory plants (for fungi, see Pilz, Molina, and Amaranthus this issue), these objectives might be met by studies that would evaluate the relationship of site characteristics to plant establishment and growth, examine cultivation treatments and outplanting, or assess genetic architecture and population biology. The basic research approach–synthesis and analysis of habitat and plant interactions–should emphasize studying species in place to address questions relevant to conservation biology, large scale monitoring strategies, and ecosystem functions.

With commercially important plants, multiple values should be addressed. Thus, research also might address questions that include how harvest affects growth and reproductive processes, population dynamics, and other interactions such as producing food or habitat for other species. In a social context research might examine rural communities and how NTFPs relate to human values, or ethnobotanical research, in which a species is understood for its cultural and spiritual role as well as its traditional utility.

The second perspective is product-centered. Research in this case would address questions of economic sustainability, value, pricing, markets, management, and product-societal relationships. (Note that biological studies are not excluded from this perspective.) Product-centered studies would result in the following:

- Increased understanding of environmental and biological effects on product yield quality, quantity, and composition.
- Development of alternative cultivation strategies for production.
- Increased understanding of agroforestry potential.
- Integration of resource or supply information with market and demand information.
- Strategies for community-based economic diversification.

A comprehensive approach to the NTFP issue would incorporate both perspectives and might involve basic information gathering about the issues, defining the scope of products, players, resources, land-bases, and ownerships involved, as well as basic information about the species, its habitat, range, etc. This kind of research, covering multiple products over a variety of spatial scales, may evolve into a systems approach where observational, computational, and communications technologies are integrated, and database information systems are used to develop complex resource models (National Research Council 1995).

NTFP RESEARCH IN THE PACIFIC NORTHWEST RESEARCH STATION, USDA FOREST SERVICE

In no other region of the United States is the research potential of NTFPs being explored more than in the Pacific Northwest. Large tracts of public and private forested lands and a history of forest-based industries mark the Pacific Northwest. Thus, it is not surprising that NTFPs received public and legislative attention, raised a variety of management issues, and revealed an alarming lack of information when markets for mushrooms and other products began to expand in the 1990's. NTFPs have contributed consistently to the social and economic fabric of the forested regions of the Pacific Northwest. Yet little was known about them. Recognizing these facts, Forest Service scientists in the region began to apply their various areas of expertise to address NTFP resource issues.

Research is a major program of the Forest Service, the largest forest resource management agency in the world. The agency provides the scientific support needed to manage and sustain the natural resources of 1.6 billion acres of forests and rangelands including 191.6 million acres within the National Forest System. The Pacific Northwest (PNW) Research Station, one of seven stations within the agency, performs basic and applied research on primarily regional issues. The broad research program direction at the PNW Research Station has enabled scientists to be responsive to changes in the social environment and to the emergence of such issues as ecological sustainability and joint resource use (Mills 1997).

PNW research on NTFPs was given impetus in the early 1990s by the demand for information on the Pacific yew (*Taxus brevifolia* Nutt.), a tree that was being harvested for Taxol, an anti-cancer compound found in the bark. To develop an environmental impact statement and conservation guidelines, information about the species was crucially needed and discovered to be lacking. In order to harvest bark from trees on public lands, funding was made available for developing conservation guidelines and an environmental impact statement through agreements made between Bristol-Myers Squibb Company, BLM, and the Forest Service (Suffness and Wall 1995). Additional funding through National Cancer Institute competitive grants was available for research on yew genetics and biochemistry as it influenced targeted compounds.

The PNW Research Station produced a variety of publications that covered a range of topics on Pacific yew. Different environmental, physiological, and cultural effects were found to influence growth and the composition and content of secondary products in Pacific yew (Kelsey and Vance 1992, Vance et al. 1994). Partnerships with researchers in the USDA Agricultural Research Service and the USDA National Germplasm Storage Center helped to support *in vitro* embryo propagation techniques for tissue culture development for the production of taxol and physiological studies of seeds for cryogenic preservation (Gibson et al. 1993, Vertucci et al. 1996). In collaboration with the Forest Service's J.H. Stone Nursery, vegetative propagation and seed, handling, storage and germination techniques were developed. Conservation research included studies of vegetative regeneration (Minore et al. 1996), studies of reproductive ecology that identified environmental and biological factors affecting flowering, seed production, and availability (Difazio et al. 1996), and ecological studies that examined the species' population dynamics (Busing and Spies 1995, Busing et al. 1995). The Pacific yew tree supplied the resource base for a physician-administered and controlled chemotherapeutic agent developed by a major pharmaceutical company having to meet U.S. Food and Drug Administration regulations, requiring clinical trials, and involving inter-governmental and industry partnerships. This does not represent the traditional development of NTFPs. However, continued sourcing for biologically-based compounds by pharmaceutical companies in temperate forests increases the probability that other plants will be sought out and developed in the same way.

More typically, products harvested in the Pacific Northwest such as edible mushrooms, boughs, moss, and medicinal herbs are sold to multiple and varied buyers, distributors and processors. The products are collected on any lands that allow access and are harvested by a process called "wildcrafting," which means generally hand-collected from the wild in a way that requires special skills and knowledge. Hundreds of organisms are harvested in different ways, at different times of the year, from different populations distributed across diverse landscapes, creating a management challenge upon which are imposed a set of regulations. Thus, much research responds to the needs of managers on public lands for monitoring and management strategies that can be integrated into existing land-use policies and directions.

Most of the research on NTFPs between 1992 and the present has been carried out by biological scientists with the chief emphasis given to biological studies of commercially important edible fungi and understory plant species (Table 1). Of the 25% that were not biologically oriented, a few projects considered economic aspects and several integrated economic with ecological information. About 60% of the studies focus on single species or guilds of species. These species are important in the Pacific Northwest because of their commercial value and because of the concerns that land managers, harvesters, conservationists, and other stakeholders have about sustainability and maintenance of these valued species. Not only are the answers generated from these studies of importance to managing these species, they contribute information to understanding how many components of forest ecosystems function.

About 35% of research and related activities have been supported to some degree through outside funding. Notably, virtually 100% of the studies have received some form of in-kind support from a variety of

TABLE 1. The information is based on a summary analysis of 20 projects that were reported in response to a questionnaire sent out to PNW scientists late in 1996. The figures in percent indicate the proportion of the 20 projects that had the identified characteristics.

Percent	Types of Studies
70	Field or lab studies.
30	Database, inventory, or other form of information.
75	Biological/ecologically based.
25	Broad issue coverage, or socio-economic
25	Harvest related (as a component).
75	Ecological, conservation, monitoring, and integrated.
35	Funded in part with outside dollars.
100	Have cooperators, or in kind support.

sources including student volunteers, state, private, and federal land-management agency personnel. Many of the cooperators are from the National Forest System, other land-management agencies such as Bureau of Land Management, and state land managers. Partnerships with industry are few, as many of those interacting most directly with the resource are frequently poorly capitalized and have little time or resources to spare. Nevertheless, projects have been carried out with the help of enthusiastic in-kind support and have generated an array of information. In-kind support takes on other forms such as donation of lab space and equipment, protected areas to carry out research, information gathering, etc. This suggests that the research is characterized by strong partnerships, is responsive to client issues, and is being conducted closely with a variety of stakeholders.

Another important factor that characterizes NTFP research activities is the diversity of objectives being met and the integration of information into broader PNW issues. The 25% directed to broad issue coverage is, in part, a response to the vast need for basic information and databases that can give the "big picture." The activities are well distributed among resource conservation research, ecosystem monitoring, basic information gathering, and education or information sharing. This corresponds to a similar distribution among the information outlets which range from refereed publications to community college courses and workshops.

From 1992 to 1997, nearly 120 publications, papers and other educational products such as workshops, community college courses, contributions to public meetings, posters, and other non-published papers were produced by PNW research programs (Table 2). The balance between publications and other kinds of information and technology transfer suggests that NTFP issues and problems yield a range of research approaches, from basic science to more issue-directed studies and analyses with components that have immediate application and use.

Most NTFP products from the PNW Research Station are biological-science based, reflecting the greater representation of biological and physical scientists at the Station. This influences the kind of information produced, and who the users of this information would be. Nevertheless, through collaborative agreements research has begun to address the social aspects of NTFPs in the Pacific Northwest. Social scientists across the United States are beginning to examine NTFPs in the context of non-traditional economies and community-based strategies for economic development (similar to research that has been reported for some time abroad (Freese 1997, Clay 1996). Collaborations with social scientists outside the Forest Service should produce information complementary to that produced by agency biologists. Since this study, economists and social scientists at the PNW Research Station have also begun studies addressing how new community-based enterprises and diverse social and land management issues evolve together.

TABLE 2. Analysis of NTFP research outputs at the Pacific Northwest Research Station from 1992-1997. Tabulation of publications, papers, posters and other technology transfer by research programs involved in NTFP research.

Program Type	Publications		Non-published papers, posters		Other technology transfer	
	No.	Percent	No.	Percent	No.	Percent
RMP*	17	34	18	69	16	36
EP	14	29	2	8	26	59
SEV	2	4	1	4		
PFHP	16	33	5	19	2	5
Total	49	100	26	100	44	100

*RMP, Resource Management and Productivity; EP, Ecosystem Processes; SEV, Social and Economic Values; PFHP, Protection of Forest Health and Productivity. PFHP is now part of Managing Natural Disturbances to Maintain Forest Health Program.

Research may be especially valuable if it provides fundamental information on multiple aspects of an issue and disseminates it through a variety of vehicles from seminars to on-line databases. Several new projects focus on developing NTFP databases that interface with databases such as the web-based USDA Natural Resource Conservation Service (NRCS) PLANTS National Database (http://plants.usda.gov/plants). Also, general information on the sustainability and conservation of commercial species in the Pacific Northwest is being developed through partnership with a non-governmental organization and funding from the State and Private Forestry branch of the Forest Service. Integrating autonomous databases so that ecological and biological information can be linked to other information on NTFP species is possible through coalitions such as the Oregon Coalition for Interdisciplinary Databases (OCID, http://www.nacse.ort/ocid), formed in 1995. Database-housed information will probably play a greater role for the next generation in providing knowledge, technology and applications to enhance competitiveness in NTFPs and other natural resource-based enterprises and meet the problems of changing values in conservation and stewardship of natural resources.

Forest Service research has been known for its strong application with specific product-directed goals and client-based outcomes. This seems to be especially true for developing research on NTFPs. The study described here, although representing a single research unit, demonstrates the strength of the Forest Service's research branch in developing information on NTFPs, aided by its effective collaborations and partnerships.

REFERENCES

Busing, R.T., C.B. Halpern, and T.A. Spies. 1995. Ecology of Pacific yew (*Taxus brevifolia*) in western Oregon and Washington. Conservation Biology 9(5):1199-1207.

Busing, R.T., and T.A. Spies. 1995. Modeling the population dynamics of Pacific yew. Research Note. PNW-RN-515. U.S. Department of Agriculture, Forest Service, Pacific Northwest Research Station, Portland, OR.

Clay, J.W. 1996. Generating income and conserving resources: 20 lessons from the field. World Wildlife Fund Publications, Baltimore, MD.

Committee for Economic Development. 1998. America's basic research-prosperity through discovery. Research and Policy Committee. Committee for Economic Development, New York, NY.

DiFazio, S.P., M.V. Wilson and N.C. Vance. 1996. Factors limiting seed production in *Taxus brevifolia* (Taxaceae) in western Oregon. Canadian Journal of Botany 74:1943-1946.

Gibson, D.M., R.E.B. Ketchum, N.C. Vance, and A.A. Christen. 1993. Initiation and growth of cell lines of *Taxus brevifolia* (Pacific yew). Plant Cell Rep. 12: 479-482.

Freese, C. 1997. The commercial, consumptive use of wild species: managing it for the benefit of biodiversity. TRAFFIC USA/World Wildlife Fund, Washington, DC.

Jungwirth, L. and B.A. Brown. 1997. Special forest products in a forest community strategy and co-management schemes addressing multicultural conflicts. In: pp. 88-107. N.C. Vance and J. Thomas (eds.). Special forest products-biodiversity meets the marketplace. GTR-WO-63, U.S. Department of Agriculture, Forest Service, Washington, DC.

Kelsey, R.G. and N.C. Vance. 1992. Taxol and cephalomannine concentrations in the foliage and bark of shade-grown and sun-exposed *Taxus brevifolia* trees. Journal of Natural Products 5:912-917.

Mills, T.J. 1997. Research priorities for entering the 21st century. Miscellaneous Report, January. U.S. Department of Agriculture, Forest Service, Pacific Northwest Research Station, Portland, OR.

Mills, T.J., F. H. Everest, P. Janik, B. Pendleton, C.G. Shaw III and D.N. Swanston. 1998. Science-management collaboration: lessons from the revision of the Tongass National Forest plan. Western J. Applied Forestry 13: 90-96.

Minore, D. and H.G. Weatherly. 1996. Stump sprouting of Pacific yew. PNW-GTR-378. U.S. Department of Agriculture, Forest Service, Pacific Northwest Research Station. Portland, OR.

Molina, R., N. Vance, J.F. Weigand, D. Pilz and M.P. Amaranthus. 1997. Special forest products: integrating social, economic, and biological considerations into ecosystem management. In: pp.315-336. K.A. Kohm and J.F. Franklin (eds.). Creating a forestry for the 21st century. Island Press, Washington, DC.

National Research Council. 1995. Finding the forest in the trees, the challenge of combining diverse environmental data. Committee for a Pilot Study on Database Interfaces, U.S. National Committee for CODATA, Commission on Physical Sciences, Mathematics, and Applications. National Academy Press, Washington, DC.

National Research Council. 1998. Forested landscapes in perspective–prospects and opportunities for sustainable management of America's nonfederal forests. National Academy Press, Washington, DC.

Pilz, D., R. Molina and M.P. Amaranthus. 2001. Productivity and Sustainable Harvest of Edible Forest Mushrooms: Current Biological Research and New Directions in Federal Monitoring. In: Emery, M.R. and R.J. McLain (eds.). Non-Timber Forest Products in the United States: Research and Policy Issues in the Pacific Northwest and Upper Midwest. The Haworth Press, Inc., New York.

Shaw, E.W. 1949. Minor Forest Products of the Pacific Northwest. Research Notes No. 59. USDA Forest Service, Pacific Northwest Forest and Range Experiment Station. Portland, OR.

Suffness, M. and M.E. Wall. 1995. Discovery and development of taxol. In: pp. 3-26. M. Suffness (ed.). Taxol® science and applications. CRC Press, New York.

Vance, N.C., R.G. Kelsey and T.E. Sabin. 1994. Seasonal and tissue variation in taxane concentrations of *Taxus brevifolia*. Phytochemistry 36:1241-44.

Vance, N.C. 1997. The challenge of increasing human demands on natural systems. In: pp. 2-7. N.C. Vance and J. Thomas (eds.). Special forest products-biodiversity meets the marketplace. GTR-WO-63, U.S. Department of Agriculture, Forest Service, Washington, DC.

Vertucci, C.W., J. Crane and N.C. Vance. 1996. Physiology of *Taxus brevifolia* seeds in relation to seed storage characteristics. Physiologia Plantarum 98:1-12.

Productivity and Sustainable Harvest of Edible Forest Mushrooms: Current Biological Research and New Directions in Federal Monitoring

David Pilz
Randy Molina
Michael P. Amaranthus

SUMMARY. The commercial harvest of wild edible forest mushrooms has increased dramatically in the Pacific Northwest United States during the last decade, creating public and managerial concerns about potential over-harvesting. These concerns have prompted Federal land management agencies and research organizations to undertake a variety of research projects addressing the ecological impacts and long-term sustainability of widespread harvesting. This article lists and briefly describes 25 ongoing research projects investigating the three most important forest mushroom genera of commerce; matsutake, morels, and chanterelles. We finish by describing future Federal directions in regional research and monitoring designed to ensure sustainable harvests through long-term cooperative monitoring involving multiple

---

David Pilz And Randy Molina are researchers with the Pacific Northwest Research Station, U.S. Department of Agriculture, Forest Service, Forestry Sciences Laboratory, 3200 SW Jefferson Way, Corvallis, OR 97331.

Michael P. Amaranthus is Associate Professor (Courtesy) with the Department of Forest Science, Oregon State University, Corvallis, OR 97331.

[Haworth co-indexing entry note]: "Productivity and Sustainable Harvest of Edible Forest Mushrooms: Current Biological Research and New Directions in Federal Monitoring." Pilz, David, Randy Molina, and Michael P. Amaranthus. Co-published simultaneously in *Journal of Sustainable Forestry* (Food Products Press, an imprint of The Haworth Press, Inc.) Vol. 13, No. 3/4, 2001, pp. 83-94; and: *Non-Timber Forest Products: Medicinal Herbs, Fungi, Edible Fruits and Nuts, and Other Natural Products from the Forest* (ed: Marla R. Emery, and Rebecca J. McLain) Food Products Press, an imprint of The Haworth Press, Inc., 2001, pp. 83-94. Single or multiple copies of this article are available for a fee from The Haworth Document Delivery Service [1-800-342-9678, 9:00 a.m. - 5:00 p.m. (EST). E-mail address: getinfo@haworthpressinc.com].

stakeholders, especially interested publics.

KEYWORDS. Mushrooms, sustainable harvest, research, monitoring

CONTEXT

The Pacific Northwest (a geographical region encompassing southeast Alaska, British Columbia, Washington, Oregon, Idaho, western Montana, and northern California) produces a diversity and abundance of wild edible forest mushrooms. Commercial harvesting of these species for local, regional, national, and international markets has grown rapidly in the last decade (Molina et al. 1993). This dramatic increase has heightened concerns about harvest sustainability, prompted land managers to examine the impact of forest management activities on mushroom crops, and stimulated research on edible mushroom biology and ecology (Pilz and Molina 1996).

A large portion of appropriate forest habitat, especially where the public is allowed to commercially collect mushrooms, is located on lands administered by two Federal agencies, the Forest Service (U.S. Department of Agriculture) and the Bureau of Land Management (U.S. Department of the Interior). During the last decade, these Federal agencies have shifted their management focus from an emphasis on timber production to managing ecosystems for a broader range of amenities and products. Resource monitoring (including edible mushrooms) is an important part of the new ecosystem management paradigm (Bormann et al. 1994).

CURRENT RESEARCH

Many individuals and organizations have an interest or "stake" in the recreational and commercial harvest of edible mushrooms from the forests of the Pacific Northwest; we refer to them collectively as "stakeholders." In addition to Federal agencies, Native American tribes, state agencies, counties, industrial timber companies, and small woodlot owners also manage forests with edible mushroom crops. Mycological societies and "personal use" mushroom harvesters abound in the region. Tens of thousands of harvesters supplement their incomes by selling the mushrooms they collect (Schlosser and Blatner 1995). Businesses in rural communities benefit from the local purchases of traveling harvesters. Mushroom buying, processing, and brokerage companies depend on this seasonal harvesting for their raw product. And of course, customers around the world relish the culinary delights of this unique forest product.

Given the large and varied public interest in wild mushroom harvesting, it is not surprising that research activities are equally numerous and diverse. Currently, biological research is concentrating on the three most valuable and widely collected edible forest mushrooms: American matsutake (*Tricholoma magnivelare* (Peck) Redhead), morels (*Morchella* species Persoon:Fr.), and chanterelles (*Cantharellus* species Fr.). Many other species are commercially collected and also deserve research and monitoring efforts, especially the king bolete (*Boletus edulis* Bull:Fr.) and truffles (*Tuber gibbosum* Harkness and *Leucangium carthusianum* [Tulasne & Tulasne] Paoletti). Table 1 summarizes current biological research activities in the Pacific Northwest. Many of these projects are sponsored or supported by different National Forests and the Pacific Northwest Research Station (management and research branches of the USDA Forest Service, respectively), but significant research also is being conducted by state universities and mycological societies. Collaboration among land management agencies, research institutions, non-government organizations, researchers, managers, harvesters, and other stakeholders has greatly broadened the scope and enhanced the quality of these research projects.

Lead Scientist Affiliations

1. U.S. Department of Agriculture, Forest Service, Pacific Northwest Research Station, Corvallis, OR 97331
2. Department of Forest Science, Oregon State University, Corvallis, OR 97331
3. U.S. Department of Agriculture, Forest Service, Winema National Forest, Chemult, OR 97731
4. U.S. Department of Agriculture, Forest Service, Region 5, Sacramento, CA 95814

5. Department of Biological Sciences, Central Washington University, Ellensburg, WA 98926
6. Biology Department, Portland State University, Portland, OR 97207
7. U.S. Department of Agriculture, Forest Service, Pacific Northwest Research Station, La Grande, OR 97850
8. Washington State Department of Natural Resources, Olympia, WA 98504
9. Oregon Mycological Society, c/o 13716 SE Oatfield Rd., Milwaukee, OR 97222
10. U.S. Department of the Interior, US Geological Survey, Biological Resources Division, Seattle, WA 98115
11. Department of Botany, University of Washington, Seattle, WA 98195
12. U.S. Department of Agriculture, Forest Service, Pacific Northwest Research Station, Olympia, WA 98512
13. U.S. Department of Agriculture, Forest Service, Wallowa Whitman National Forest, Baker City, OR 97814
14. U.S. Department of Agriculture, Forest Service, Pacific Northwest Research Station, Portland, OR 97205

FEDERAL MONITORING PLANS

Concern about the sustainability of large-scale commercial mushroom harvesting in the Pacific Northwest is partly based on declining crops in traditionally harvested areas of Europe and Japan (Arnolds 1991, Arnolds 1995, Hosford et al. 1997). Air pollution, climate change, intensive forest management, introduced forest diseases, loss of forest habitat, and intensive mushroom harvesting all have the potential to affect long-term production of edible mushrooms in our forests. Monitoring (baseline inventories repeated over time) is essential to detect trends and correlate productivity with habitat, forest management activities, and environmental conditions.

Investigating efficient sampling methods has been an important part of many of our productivity studies (last section of Table 1). These methods have recently been field-tested and manuscripts reporting our results and suggested protocols are in preparation. We have also designed a regional edible mushroom monitoring program that will address the need for cost-effective monitoring. Although details are still

TABLE 1. Edible forest mushrooms: biological research and monitoring activities in Oregon and Washington

Research Project	Lead Scientists	Description
American matsutake		
Matsutake production and value in forest habitats of Oregon (Pilz et al. 1999)	Pilz[1], Amaranthus[2], Molina[1], Luoma[2]	Determines matsutake production and value from long-term study plots in known matsutake fruiting areas in the Winema, Siskiyou, Umpqua, and Siuslaw National Forests.
Economic comparisons of matsutake and timber values in high elevation forests on the eastern slopes of the Cascade Range in southern Oregon (Pilz et al. 1999)	Pilz, Alexander[1], Brown[3], Molina	Describes considerations for comparing the present net worth of matsutake and timber produced on the same land under differing timber management matsutake habitat enhancement scenarios. Illustrates how to conduct the comparison with examples.
Effect of harvest and soil raking on matsutake production and value in forest habitats of Oregon (Pilz et al. 1996a)	Amaranthus, Pilz, Luoma, Molina	Determines matsutake production and value after mushroom harvest and various intensities of soil raking and duff replacement in several Oregon habitats.
American matsutake production across spatial and temporal scales (Amaranthus and Powers 1999)	Amaranthus, Pilz, Luoma	Evaluates production and value of matsutake mushrooms over temporal scales. Spatial relationships to various individual trees and habitats is assessed.
Management experiments for developing agro-forestry systems producing matsutake mushrooms (Weigand 1998)	Weigand[4], Amaranthus	Compares experimental prescriptions for developing and refining agro-forestry systems that increase financial returns via matsutake production in the Cascade Range in southern Oregon. Examines management practices to increase matsutake fruiting, including manipulating leaf area, density, composition, structure, and forest floor litter depth.
Shiro analysis of matsutake in the Cascade Range in central Washington (Hosford et al 1997)	Hosford[5]	Examines matsutake production, seasonal variation in sporocarp location, maturation interval, plant community, and soil conditions across a variety of shiro locations.
American matsutake ectomycorrhizal morphology and genetics	Lefevre[2], Molina	Examines mycorrhizal host range of American matsutake with various arboreal host species, describes mycorrhizae, and conducts genetic analysis of symbiosis with the achlorophyllous plant *Allotropa virgata*.
Genetic variability of American matsutake populations	Carter[6], Lefevre, Molina	Develops molecular probes and conducts PCR-based analysis of population variability of American matsutake from sites around the Pacific Northwest.
Protocols for sampling matsutake mycelial mats for estimating site occupancy.	Lefevre, Molina	Develops sampling methods for field identification of matsutake mycelium in soil cores using visual, olfactory, and textural criteria. Applies sampling methods to estimates of matsutake site occupancy and abundance.

TABLE 1 (continued)

Research Project	Lead Scientists	Description
Morels		
Morel productivity, ecology, and population genetics after wildfire and tree mortality in northeastern Oregon	Pilz, Weber[2], Parks[7], Carter, Molina	Investigates morel productivity, ecology, taxonomy and population genetics after wildfire and tree mortality in mixed-conifer stands of northeastern Oregon.
Morel production and species composition after fuel reduction treatments in northeast Oregon	Smith[1], Weber, Carter, Molina	Compares morel production after no treatment, forest thinning, forest thinning with prescribed fire, and prescribed fire only. Examines genetic variability and taxonomic distinctions among morel populations.
Morel productivity, taxonomy, and population genetics in south-central Oregon (Weber et al. 1996)	Weber, Pilz, Carter, Molina	Tests selected ways of quantifying information on abundance; develops formula for predicting sporocarp weight and size; gathers samples for morphological, taxonomic and population genetic studies.
Morel productivity in mixed conifer and Pacific madrone stands after wildfire in southwestern Oregon	Amaranthus	Compares morel production in no fire, light, medium, and high fire intensity areas in madrone and mixed-conifer plots after the Ramsey Wildfire near Gold Hill, Oregon.
Morel mycorrhizae (Dahlstrom et al. In Prep.)	Smith, Weber, Dahlstrom[2], Fujimura[2], Barroetavena[2], Molina	Investigates morphological and physiological characteristics of putative mycorrhizal symbioses between *Morchella* species and arboreal hosts of the Pacific Northwest.
Chanterelles		
Biological inventory of chanterelle productivity in various forest types on the Olympic Peninsula (Pilz et al. 1998a)	Pilz, Molina, Liegel[1]	Examines sampling methods, productivity per unit area, productivity by commercial grade, mushrooms size/weight correlations for two years in various forest habitats. Part of a larger socio-economic and managerial study sponsored by the United Nations Man and the Biosphere (MAB) Program. Many cooperators, including the Puget Sound Mycological Society, various landowners, and commercial mushroom buyers.
Economic comparisons of chanterelle and timber values on the Olympic Peninsula (Pilz et al. 1998b)	Pilz, Brodie[2], Alexander[1], Molina	Describes considerations for comparing the present net worth of chanterelles and timber produced on the same land under differing timber management scenarios. Illustrates how to conduct the comparison with examples.

Effect of forest and mushroom harvest on chanterelle production on the Olympic Peninsula (Amaranthus and Russel 1996)	Amaranthus, Bigley[8]	Evaluates silvicultural treatments and chanterelle harvest versus no harvest for effects on chanterelle fresh weight production. Examines relations between productivity and forest floor vegetation.
Oregon *Cantharellus* study project (Norvell 1995, Norvell et al. 1995)	Norvell[9], Roger[9]	Determines the effects of long-term chanterelle harvesting and harvest methods on subsequent fruiting. Assesses associated fungus and plant diversity. Correlates fruiting with weather patterns.
Chanterelle production responses to young stand thinning (Pilz et al. 1996b)	Pilz, Molina	Evaluates chanterelle production in uncut control areas and in two young forest thinning treatments (125 and 300 residual trees/hectare). Maps chanterelle fruiting of mycelial colonies relative to live and dead trees before and after logging. Examines sampling methods.
Genetic structure of chanterelle populations on the Olympic Peninsula (Rodriguez et al. In Prep.)	Rodriguez[10], Redman[10], Ammirati[11]	Develops a PCR-based system to measure temporal and spatial variation in genetic diversity of populations. Evaluates the impact of harvesting sporocarps on genetic structure. Determines the genetic structure of isolates associated with plant roots versus sporocarps. Program is entering the fourth year of a ten year study.
Community structure and dynamics of ectomycorrhizal fungi in managed forest Stands (Luoma et al. 1996)	Luoma	Compares response of chanterelle production to different levels of green-tree retention as part of the Demonstration of Ecosystem Management Options (DEMO) research project.
Variable density thinning; forest ecosystem study (Cary et al. 1999)	Carey[12], Thysell[12]	Examines the effect of young stand thinning on mycorrhizal and non-mycorrhizal fungal diversity and productivity. Sampled fungi include chanterelles and other edible mushrooms. This study is a part of a larger forest ecosystem study investigating methods to speed the development of old-growth stand characteristics in young stands.
Species distinctions, population genetics and habitat specificity of commercially harvested yellow *Cantharellus* species	Dunham[2], O'Dell[1], Molina	Explores species distinctions among several look-alike yellow *Cantharellus* species that are commercially harvested, examines habitat relations of each species, and examines stand-level population genetics of *Cantharellus formosus* genets.

TABLE 1 (continued)

Research Project	Lead Scientists	Description
Modeling the occurrence and distribution of 9 cantharelloid fungi from 3 genera (6 are commercially-harvested edible species)	Dreisbach[1], Smith[1], Molina	Develops predictive models using Habitat Suitability Indices (HSI) applied to GIS layers to delineate landscape-level distribution and quality of forest habitat for related canterelloid fungi with widely-varied specificity in habitat requirements. Uses these models to evaluate Federal forest plans for land allocations and management activities needed to create and maintain appropriate forest habitat to sustain suites of organisms dependant on late-successional forests.
Economics and Monitoring		
Price projecttions of commercial mushrooms and timber in the Pacific Northwest (Alexander et al. In Prep.)	Alexander, Pilz, Brown, Rockwell[13]	Illustrates how to compare the present net worth of various commercial mushroom species and timber produced on the same land under differing timber management scenarios.
Comparison of forest mushroom sampling methods	Pilz, Max[14], Huso[2], Molina	Statistically compares the efficiency of various sample plot sizes, shapes, and numbers for inventorying spatially and temporally clustered mushroom production. Data derived from 5 productivity studies.
Field trials of selected mushroom sampling protocols	Pilz, Max, Huso, Molina	Conducts field tests of sampling methods for morel, chanterelle, and matsutake mushrooms. Examines plot shape and size, sampling time, and means of reducing detection error. Emphasizes practicality under a variety of field conditions.
Regional edible mushroom monitoring (Pilz and Molina 1998)	Pilz, Molina	Proposes a long-term, three-component, regional monitoring plan for edible mushrooms. (See Federal monitoring plans section.)

in early stages of development, Federal agencies hope to involve all interested stakeholders in designing, implementing, and maintaining the monitoring program.

As currently envisioned, we propose a three-component approach to regional monitoring (Pilz and Molina 1998). One component would consist of inventory plots in natural areas where neither mushroom nor timber harvesting is allowed. This branch of the program would provide control sites for interpreting trends in mushroom productivity on monitoring sites where timber or mushroom harvesting does occur. It also would detect changes in mushroom fruiting related to dispersed regional influences on forest health such as atmospheric pollution or climate change. We propose enlisting the help of volunteers from mushroom societies for some of the field work.

Measuring total harvest quantities from areas that experience intensive commercial mushroom harvesting would constitute the second component. This branch of the program would provide critical information about whether commercial harvesting directly impacts subsequent or long-term mushroom production. We envision cooperating with commercial harvesters to collect this data in exchange for providing them with exclusive access to selected areas.

Mushroom production differs widely from year to year, from area to area, and among species. The first two program components will require decades of sampling representative sites (for each species) throughout the region, but relatively few sites will be needed and the annual effort at each site will be modest.

The third component of this monitoring program will employ statistically rigorous sampling of numerous sites to model the relations between mushroom productivity and forest habitat (stand conditions and forest management activities). We anticipate these predictive models will be applicable across the broad range of ecological, geological, physiographic, and climatic conditions that exist in the Pacific Northwest allowing forest managers throughout the region to anticipate the influence of their management activities on subsequent mushroom crops.

Lastly, this proposed monitoring program can also be characterized as having three components in another sense; by involving researchers, land managers, and the interested public in developing and implementing the monitoring system. Collaborative efforts among these three groups may serve as an example for how multiple stakeholders can be involved in the management of resources on public lands. As such, it will illustrate an essential concept of ecosystem management, that humans are an integral part of ecosystems, and wise management choices require the involvement and support of the public.

REFERENCES

Alexander, S., D. Pilz, et al. [In preparation]. Price projections of commercial mushrooms and timber in the Pacific Northwest.

Amaranthus, M. and K. Russell. 1996. Effect of environmental factors and timber and mushroom harvest on *Cantharellus* production on the Olympic Peninsula, Washington State. In pp. 73-74. Pilz, D., Molina, R. (eds.). Managing forest ecosystems to conserve fungus diversity and sustain wild mushroom harvests. Gen. Tech. Rep. PNW-GTR-371. U.S. Department of Agriculture, Forest Service, Pacific Northwest Research Station, Portland, OR.

Amaranthus, M.P., D. Pilz, A. Moore, R. Abbott and D.L. Luoma. 2000. American matsutake production across spatial and temporal scales. In Power, R.F., D.L. Hauxwell, and G.M. Nakamura (tech. Cords.). Proceedings of the California Forest Soils Council conference on forest soils biology and forest management; February 23-24, 1996; Sacramento, CA. Gen. Tech. Rep. PSW-GTR-178. U.S. Department of Agriculture, Forest Service. Pacific Southwest Research Station, Albany, CA.

Arnolds, E. 1991. Decline of ectomycorrhizal fungi in Europe. Agriculture, Ecosystems, and Environment 35: 209-244.

Arnolds, E. 1995. Conservation and management of natural populations of edible fungi. Canadian Journal of Botany 73(1): 987-998

Bormann, B.T., P.G. Cunningham, M.H. Brookes, V.W. Manning and M.W. Collopy. 1994. Adaptive ecosystem management in the Pacific Northwest. Gen. Tech. Rep. PNW-GTR-341. U.S. Department of Agriculture, Forest Service, Pacific Northwest Research Station, Portland, OR.

Carey, A.B., D.R. Thysell and A.W. Brodie. 1999. The forest ecosystem study: background, rationale, implementation, baseline conditions, and silvicultural assessment. Gen. Tech. Rep. PNW-GTR-457. U.S. Department of Agriculture, Forest Service, Pacific Northwest Research Station, Portland, OR.

Dahlstrom, J.L., J.E. Smith and N.S. Weber. [In Preparation]. Formation of morel mycorrhizae with species of Pinaceae in pure culture synthesis. [Mycorrhizae].

Hosford, D., D. Pilz, R. Molina and M.P. Amaranthus. 1997. Ecology and management of the commercially harvested American matsutake mushroom. Gen. Tech. Rep. PNW-GTR-412. U.S. Department of Agriculture, Forest Service, Pacific Northwest Research Station, Portland, OR.

Luoma, D.L., J.L. Eberhart and M.P. Amaranthus. 1996. Community structure and dynamics of ectomycorrhizal fungi in managed forest stands: demonstration of ecosystem management options (DEMO). In pp. 27-31. Pilz, D., Molina, R. (eds.). Managing forest ecosystems to conserve fungus diversity and sustain wild mushroom harvests. Gen. Tech. Rep. PNW-GTR-371. U.S. Department of Agriculture, Forest Service, Pacific Northwest Research Station, Portland, OR.

Molina, R., T. O'Dell, D. Luoma, M. Amaranthus, M. Castellano and K. Russell. 1993. Biology, ecology, and social aspects of wild edible mushrooms in the forests of the Pacific Northwest: a preface to managing commercial harvest. Gen. Tech. Rep. PNW-GTR-309. U.S. Department of Agriculture, Forest Service. Pacific Northwest Research Station, Portland, OR.

Norvell, L. 1995. Loving the chanterelle to death? The ten-year Oregon chanterelle project. McIlvanea 12(1): 6-25.

Norvell, L., F. Kopecky, J. Lindgren and J. Roger. 1995. The chanterelle (*Cantharellus cibarius*): a peek at productivity. In pp. 117-128. Schnepf, C. (comp. ed.). Dancing with the elephant: Proceedings: The business and science of special forest products–a conference and exposition; 1994 Jan. 26-27. University of Idaho Extension Service, Hillsboro, OR.

Pilz, D. and R. Molina. 1996. Managing forest ecosystems to conserve fungus diversity and sustain wild mushroom harvests. Gen. Tech. Rep. PNW-GTR-371. U.S. Department of Agriculture, Forest Service, Pacific Northwest Research Station, Portland, OR.

Pilz, D., R. Molina and J. Mayo. 1996a. Matsutake inventories and harvesting impacts in the Oregon Dunes National Recreation Area. In pp. 78-80. Pilz, D., Molina, R. (eds.). Managing forest ecosystems to conserve fungus diversity and sustain wild mushroom harvests. Gen. Tech. Rep. PNW-GTR-371. U.S. Department of Agriculture, Forest Service, Pacific Northwest Research Station, Portland, OR.

Pilz, D., R. Molina and J. Mayo. 1996b. Chanterelle production responses to stand thinning. In pp. 81-82. Pilz, D., Molina, R. (eds.). Managing forest ecosystems to conserve fungus diversity and sustain wild mushroom harvests. Gen. Tech. Rep. PNW-GTR-371. U.S. Department of Agriculture, Forest Service, Pacific Northwest Research Station, Portland, OR.

Pilz, D. and R. Molina. 1998. A proposal for regional monitoring of edible forest mushrooms. Mushroom. The Journal of Wild Mushrooming 60(16): 19-23.

Pilz, D., R. Molina and L. Liegel. 1998a. Biological productivity of chanterelle mushrooms in and near the Olympic Peninsula Biosphere Reserve. In pp. 8-13. L.H. Liegel (comp.). The biological, socioeconomic, and managerial aspects of chanterelle mushroom harvesting: the Olympic Peninsula, Washington State, U.S.A. Ambio, A Journal of the Human Environment. Special Report Number 9, September, 1998. Royal Swedish Academy of Sciences, Stockholm, Sweden.

Pilz, D., F.D. Brodie, S. Alexander and R. Molina. 1998b. Relative value of chanterelles and timber as commercial forest products. In pp. 14-15. L.H. Liegel, (comp.) The biological, socioeconomic, and managerial aspects of chanterelle mushroom harvesting: the Olympic Peninsula, Washington State, U.S.A. Ambio, A Journal of the Human Environment. Special Report Number 9, September, 1998. Royal Swedish Academy of Sciences, Stockholm, Sweden.

Pilz, D., J. Smith, M.P. Amaranthus, S. Alexander, R. Molin and D. Luoma. 1999. Managing the commercial harvest of the American matsutake and timber in the southern Oregon Cascade Range. Journal of Forestry 97(2): 8-15.

Rodriguez, R.J., J. Ranson and R.S. Redman. [In Preparation]. Genetic structure of

Cantharellus cibarius populations in a spruce/hemlock forest of the Pacific Northwest [Molecular Ecology or New Phytologist].

Schlosser, W.E. and K.A. Blatner. 1995. The wild edible mushroom industry of Washington, Oregon and Idaho: a 1992 survey. Journal of Forestry 93(3): 31-36.

Weber, N.S., D. Pilz and C. Carter. 1996. Morel life histories–beginning to address the unknowns with a case study in the Fremont National Forest near Lakeview, Oregon. In pp. 62-68. Pilz, D., Molina, R. (eds.). Managing forest ecosystems to conserve fungus diversity and sustain wild mushroom harvests. Gen. Tech. Rep. PNW-GTR-371. U.S. Department of Agriculture, Forest Service, Pacific Northwest Research Station, Portland, OR.

Weigand, J.F. 1998. Management experiments for high-elevation agroforestry systems jointly producing matsutake mushrooms and high-quality timber in the Cascade Range of southern Oregon. Gen. Tech. Rep. PNW-GTR-424. U.S. Department of Agriculture, Forest Service, Pacific Northwest Research Station, Portland, OR.

Socio-Economic Research on Non-Timber Forest Products in the Pacific Northwest

Susan J. Alexander
Rebecca J. McLain
Keith A. Blatner

SUMMARY. The non-timber forest products industry in the Pacific Northwest has been viable for nearly a century. Although it is a small part of the regional economy, the industry involves many people in the region and products are exported worldwide. Harvest of non-timber forest products has become more scrutinized as landowners, forest managers, and harvesters struggle to meet their sometimes conflicting needs and requirements, and deal with growing demand for many wild products. Much of the research on non-timber forest products has focused on biology and ecology, although there has been some research on the social and economic aspects of non-timber forest products over the past several decades. There are several social and economic studies of the industry that are underway or just being completed in the Pacific Northwest. Current research includes studies on product yield, market surveys, price analysis, product management and silviculture, recreational use, and policy analysis. Recommendations for future research

---

Susan J. Alexander is Research Economist with the USDA Forest Service Pacific Northwest Forest Sciences Laboratory, Corvallis, OR 97331 (E-mail: salexander@fs.fed.us).

Rebecca J. McLain is Co-Founder and Director of the Institute for Culture and Ecology, P.O. Box 6688, Portland, OR 97228 (E-mail: mclain@ifcae.org).

Keith A. Blatner is Professor and Chair, Department of Natural Resource Sciences, Washington State University, Pullman, WA 99164.

[Haworth co-indexing entry note]: "Socio-Economic Research on Non-Timber Forest Products in the Pacific Northwest." Alexander, Susan J., Rebecca J. McLain, and Keith A. Blatner. Co-published simultaneously in *Journal of Sustainable Forestry* (Food Products Press, an imprint of The Haworth Press, Inc.) Vol. 13, No. 3/4, 2001, pp. 95-103; and: *Non-Timber Forest Products: Medicinal Herbs, Fungi, Edible Fruits and Nuts, and Other Natural Products from the Forest* (ed: Marla R. Emery, and Rebecca J. McLain) Food Products Press, an imprint of The Haworth Press, Inc., 2001, pp. 95-103. Single or multiple copies of this article are available for a fee from The Haworth Document Delivery Service [1-800-342-9678, 9:00 a.m. - 5:00 p.m. (EST). E-mail address: getinfo@haworthpressinc.com].

are outlined. The non-timber forest product industry is a highly varied and frequently changing industry, with issues ranging from biological sustainability to equity. Social and economic research helps resolve questions surrounding management, harvesting, production and marketing of these highly demanded and often poorly understood products.

KEYWORDS. Non-timber forest products, NTFP industry, NTFP markets, NTFP socio-economics

MOTIVATIONS FOR SOCIO-ECONOMIC RESEARCH

Although much of the work in the Pacific Northwest on non-timber forest products (NTFPs) has focused on biological and ecological aspects of NTFPs, a small but growing portfolio of projects also addresses sociological and economic aspects of NTFP use and management in the Pacific Northwest. The research is supported by the USDA Forest Service Pacific Northwest Research Station, the state of Oregon, Oregon State University, the state of Washington, University of Washington, Washington State University, and individuals in the NTFP industry, among others. These organizations and individuals have multiple reasons to support such projects. Public and private land managers want to know the economic value of the NTFPs being harvested on their lands, so they have a better idea of prices to charge for permits and leases and so they can better weigh the economic benefits and costs of managing an area for NTFPs. Forest managers want to be able to determine the conditions that favor production of different NTFPs so they can better assess the impacts of different forest management options. Forest managers are interested in obtaining information about reproduction rates and yields under different harvesting regimes, which would allow them to set realistic limits on allowable quantities and seasons. Forest managers want to know who is harvesting NTFPs, how much various users harvest, what their incomes are from these activities, and how various policies affect different types of harvesters. In part this information allows managers to tailor programs to address a variety of user group needs and in part it permits them to balance equity and efficiency concerns associated with different policy and regulatory options. An overview of some of the projects in the Pacific Northwest that address social and economic issues of NTFPs is provided below.

RECENTLY COMPLETED AND ON-GOING RESEARCH PROJECTS WITH SOCIO-ECONOMIC COMPONENTS

Beargrass Production Study, Gifford Pinchot National Forest, Washington State[1]

Blatner and Higgins of Washington State University are working on a five-year study of beargrass (*Xerophyllum tenax* [Pursh] Nutt.) production in the Cispus Adaptive Management Area in the Gifford Pinchot National Forest, in western Washington. The study will provide estimates of how much harvestable beargrass is produced under different overstory conditions. It will also provide data on rates of regrowth over a five-year period.

Noble Fir Bough Yield Study, Washington and Oregon[2]

Fight, Savage, Blatner, Vance, Chapman and Cooper are conducting a noble fir (*Abies procera* Rehder) bough yield study in Washington and Oregon based on two preliminary models. The first model predicts the number of harvestable whorls per tree, while the second predicts the average weight per whorl. The study design initially covered 238 sample plots, and was expanded to 400 plots in 1997. Analyses are focusing on the utility of the models and on the development of cost effective techniques for estimating harvestable weight from existing noble fir stands in the region. Project researchers will also develop a socio-economic profile of bough harvesters.

Floral and Christmas Greens Market Survey, Pacific Northwest[3]

Blatner and Schlosser have completed a resurvey of the floral greens and the Christmas greens and ornamentals industries in the Pacific Northwest. The study is a follow-up to their survey of the floral greens industry in 1989. Preliminary results indicate that the Christmas bough and ornamental market has continued to grow since 1989, while the floral greens market has remained constant.

NTFP Price Study, Pacific Northwest[4]

Blatner and Alexander have published the results of a price study of NTFP prices for the states of Washington, Oregon and Idaho (Blatner and Alexander 1998). They reported on prices for floral greens, Christmas greens, wild edible mushrooms, and wild edible berries. One of the initial goals of the study was to interest the Northwest Special Forest Products Association, whose membership is composed of NTFP buyers and dealers, in continuing the price reporting on a regular basis. It is unclear whether the Association will be able to continue the research.

Markets and Silviculture of Pacific Northwest Matsutake, Pacific Northwest[5]

Weigand has recently completed a two-year study on Pacific Northwest American matsutake (*Tricholoma magnivelare* [Peck] Redhead). His research describes the international market and assesses export prices and price elasticities for this commercial species. He also describes the silviculture of American matsutake in the Pacific Northwest. For summaries of this work, refer to Amaranthus (1998) and Weigand (1998a, 1998b).

Matsutake Mushroom Study, Oregon[6]

Pilz, Amaranthus, Smith, Molina, Moore and others have studied American matsutake production in Oregon for several years. The data gathered on production and prices was used to analyze productivity rates, American matsutake values, and the effects of harvesting techniques on productivity levels. The production and biology of American matsutake, with a brief discussion of economics, was recently published in the *Journal of Forestry* (Pilz et al. 1999). Information gathered in the study is being used for other analyses, such as an estimate of wild mushroom value in different regions of the Pacific Northwest.

Recreational Huckleberry and Mushroom Picker Study, Gifford Pinchot National Forest, Washington[7]

Alexander, Kim and Johnson are analyzing survey response data in a study of recreational huckleberry and mushroom pickers on the Gifford Pinchot National Forest. The research objective is to estimate nonmarket prices for huckleberries and for some species of wild edible mushrooms, and to assess survey respondent comments about the resources.

U.S. Man and Biosphere Wild Chanterelle Study; Sociological Component, Olympic Peninsula, Washington[8]

In 1996 Love and Jones (1997) completed a three-year study of chanterelle (*Cantharellus* species) mushroom harvesters on the Olympic Peninsula. They identified key socio-economic characteristics of chanterelle harvesters in the study area, existing use of stewardship practices, and documented the tensions that exist between scientists, managers, and chanterelle harvesters and buyers in the study area. A subsequent sociological analysis of the project carried out by McLain and Jones (1996) describes and analyzes the barriers and opportunities for carrying out interdisciplinary and multi-stakeholder research projects on NTFP issues in the Pacific Northwest. The results of the entire series of studies have been published as an Ambio special report.[9]

Knowlege, Rules, and Policy Development, Olympic Peninsula[10]

In 1994, Robinson (1994) and Kantor (1994) completed an exploratory study that examined knowledge networks among NTFP stakeholders and the role of NTFP harvesters and buyers in the development of rules and regulations governing NTFP harvesting on the Olympic Peninsula. As part of the same project, McLain (1998) examined the role of amateur scientists in NTFP politics in the Pacific Northwest.

Wild Mushroom Policy Study, Oregon[11]

McLain is conducting an ethnographic study of wild mushroom politics in central Oregon. The study documents harvester and buyer strategies for coping with the uncertainties associated with wild mushroom harvesting. Her work also analyzes the effects of newly developed wild mushroom policies on harvesters and harvesting activities and describes the roles that harvesters currently play in policy decisions that affect their access to wild mushroom grounds on national forests in central Oregon.

Wild Mushroom and Timber Values in the Pacific Northwest[12]

Alexander, Pilz and others are developing estimates of present net value for both timber and mushrooms, when both are produced from the same land base, for three commercial mushroom species in a variety of habitats in Oregon and Washington. Such estimates are, by necessity, specific to particular habitat types and management activities. The values give a good idea of the scope and variety of values for both timber and wild mushrooms that can be produced from a wide range of timber types.

FUTURE RESEARCH NEEDS AND CONCERNS

Future research on social and economic aspects of NTFPs in the Pacific Northwest needs to focus on the following areas.

1. Development of information about and models for estimating harvestable quantities of NTFPs is critical to the business and management of NTFPs. Such models need to be structured to enable analysts to examine sustainable harvest issues related to the growth, function, management and ecology of commercial NTFPs. Given the perceived increasing demand for NTFP and the lack of knowledge about inventories, sustainability, or even realistic estimates of demand and use, information in this area is critically needed to balance public demand with resource sustainability.

2. A regular collection of price data for the major NTFP commercial species would be useful to harvesters, buyers, land managers, insurance companies, and others. Regular price data would be useful in developing an understanding of seasonal variation and long-term trends. For species with volatile prices, knowledge of how prices fluctuate over time would allow land managers to set and adjust permit and lease fees with sensitivity to shifting market values. Long-term

price data would also lead to greater accuracy in determining prices used in calculating NTFP net present values on the basis of biological yield.

3. Studies of NTFP gatherers' household economies, knowledge, and stewardship practices would have a variety of applications. Household economic studies would provide scientists and land managers with a better understanding of the role that NTFPs play in overall livelihood strategies for rural and urban residents. Studies of gatherer knowledge and stewardship practices could help inform ecologically sensible management and the information could be disseminated to harvesters.

4. It is important to identify economic, social, and political barriers to and opportunities for harvester and buyer participation in public decision-making forums and scientific research projects. Such information would allow land managers to develop approaches to decision-making that are more accessible to harvesters and buyers, thus increasing the likelihood that important information about NTFPs and the people who harvest, buy and process them is not left out.

5. There is a need for the development of research methodologies and approaches that are structured in ways that fit better with the agendas and working styles of NTFP businesses. For example, effective price reporting is dependent on the cooperation of NTFP processors. The high turnover in NTFP companies makes it extremely difficult to maintain an up-to-date listing of NTFP processing companies and for researchers to maintain the level of personal rapport with the processors needed to regularly collect this type of information using survey approaches. Moreover, there is opposition to periodic annual surveys among NTFP processors in the Pacific Northwest. To overcome these obstacles, some industry members and scientists have suggested that the industry make an effort to generate price date on its own. The prospects that this will happen, however, are slim since many NTFP business owners are not convinced that such data collection is worthwhile. An alternative that has been suggested by anthropologists working on NTFP issues in the Pacific Northwest is that greater use be made of participant observation and participatory research to gather economic and marketing data rather than relying on the more traditional survey methodologies that have been used up until now.

6. A greater integration of social science and economic research with on-going biological research programs and projects in the Pacific Northwest region is important.

7. Finally, the development of a more coordinated program of social science and economic research that links existing projects within the Pacific Northwest Region to each other and that establishes links to related research projects being carried out in other areas of the United States, Canada and Mexico is needed. Greater coordination of research on NTFPs within regions, ecozones, and across countries would allow scientists to conduct comparative studies, suitable for identifying broad economic and social processes that constrain or facilitate the development of particular forms of NTFP harvesting and management systems.

The NTFP industry is constantly changing; examples of current change are frequent entry and exit of firms, increasing vertical integration of individual firms, issues about land tenure and harvesting rights, and the identity of harvesters. Landowners and industry people need information about many issues to enable them to manage resources and public demand effectively, and remain in business. Researchers need to understand the issues, both past and present, to help landowners and industry participants balance resource sustainability and public demand. Social and economic research, both that underway and that suggested, will help all those involved in NTFP research, management, harvesting, production and marketing.

NOTES

1. Contact: Keith A. Blatner. Washington State University. Department of Natural Resource Science. Pullman, WA 99164 (email: dog1@cahe.wsu.edu).

2. Contact: Roger Fight. USDA Forest Service Pacific Northwest Research Station. 1221 Yamhill, Portland, OR 97205 (email: rfight@fs.fed.us).

3. Contact: Keith Blatner

4. Contact: Keith Blatner, or Susan J. Alexander. USDA Forest Service Pacific Northwest Research Station. 3200 SW Jefferson Way. Corvallis, OR 97331 (email: salexander@fs.fed.us).

5. Contact: James F. Weigand. USDA Forest Service Sierra Nevada Framework Project. 801 I Street, Sacramento, CA 95814.

6. Contact: Dave Pilz. USDA Forest Service Pacific Northwest Research Station, 3200 SW Jefferson Way, Corvallis, OR 97331 (email: dpilz@fs.fed.us).

7. Contact: Susan J. Alexander.

8. Contact: Thomas Love, Linfield College, McMinnville, Oregon (email: tlove@linfield.edu); or Eric Jones. Institute for Culture and Ecology, P.O. Box 6688, Portland, OR 97228 (email: etjones@ifcae.org).

9. Ambio special report#9, September 1998.

10. Contact: Rebecca J. McLain, Institute for Culture and Ecology, P.O. Box 6688, Portland, OR 97228 (email: mclain@ifcae.org).

11. Contact: Rebecca J. McLain.

12. Contact: Susan J. Alexander.

REFERENCES

Amaranthus, M.P., J.F. Weigand, and R. Abbott. 1998. Managing high-elevation forests to produce American matsutake (*Tricholoma magnivelare*), high-quality timber, and nontimber forest products. Western Journal of Applied Forestry 13(4): 120-128.

Blatner, K.A. and S. Alexander. 1998. Recent price trends for non-timber forest products in the Pacific Northwest. Forest Products Journal 48(10): 28-34.

Kantor, S. 1994. Local knowledge and policy development: Special forest products in coastal Washington. Master's thesis, University of Washington, Seattle, WA.

McLain, R.J., H.H. Christensen, and M.A. Shannon. 1998. When amateurs are the experts: amateur mycologists and wild mushroom politics in the Pacific Northwest, USA. Society and Natural Resources 11: 615-626.

Pilz, D., J. Smith, M.P. Amaranthis, S. Alexander, R. Molina, and D. Luoma. 1999. Managing the commercial harvest of the American matsutake and timber in the southern Oregon coast range. Journal of Forestry 97(2): 8-15.

Robinson, C.M. 1994. Multiple perspectives: Rules governing special forest product management in coastal Washington. Master's thesis, University of Washington, Seattle, WA.

Weigand, J.F. 1998. Forest management for the North American pine mushroom (*Tricholoma magnivelare* (Peck) Redhead) in the southern Cascade range. Ph.D. diss., Oregon State University, Corvallis, OR.

Weigand, J.F. 1998. Management experiments for high-elevation agroforestry systems jointly producing matsutake mushrooms and high-quality timber in the Cascade Range of southern Oregon. Gen. Tech. Rep. PNW-GTR-424. Portland, OR: USDA Forest Service, Pacific Northwest Research Station.

SECTION III: SOCIO-POLITICAL CONSIDERATIONS FOR NON-TIMBER FOREST PRODUCT MANAGEMENT

Why Is Non-Timber Forest Product Harvesting an "Issue"? Excluding Local Knowledge and the Paradigm Crisis of Temperate Forestry

Thomas Love
Eric T. Jones

SUMMARY. Despite an encouraging trend in North America of growing interest across a range of disciplines in non-timber forest products

Thomas Love is Professor of Anthropology, Linfield College, McMinnville, OR 97128.

Eric T. Jones is Co-Founder and Director of the Institute for Culture and Ecology, P.O. Box 6688, Portland, OR 97212 (E-mail: etjones@ifcae.org).

[Haworth co-indexing entry note]: "Why Is Non-Timber Forest Product Harvesting an 'Issue'? Excluding Local Knowledge and the Paradigm Crisis of Temperate Forestry." Love, Thomas, and Eric T. Jones. Co-published simultaneously in *Journal of Sustainable Forestry* (Food Products Press, an imprint of The Haworth Press, Inc.) Vol. 13, No. 3/4, 2001, pp. 105-121; and: *Non-Timber Forest Products: Medicinal Herbs, Fungi, Edible Fruits and Nuts, and Other Natural Products from the Forest* (ed: Marla R. Emery, and Rebecca J. McLain) Food Products Press, an imprint of The Haworth Press, Inc., 2001, pp. 105-121. Single or multiple copies of this article are available for a fee from The Haworth Document Delivery Service [1-800-342-9678, 9:00 a.m. - 5:00 p.m. (EST). E-mail address: getinfo@haworthpressinc.com].

(e.g., this volume), NTFP harvesters' knowledge and practices continue to be poorly understood and undervalued, if not ignored, both by research scientists and forestland policy-makers and managers. This article explores why NTFP harvesting suddenly emerged in North America as an "issue" in the early 1990s. Drawing from a three-year study of chanterelle mushroom harvesters on the Olympic Peninsula Biosphere Reserve (Washington, USA), we discuss a variety of forces which intersected in this period to bring NTFP harvesting to wider attention. Unfortunately, harvesters continue to be excluded as knowledgeable actors in, if not legitimate co-managers of, temperate forest ecosystems, resulting in both passive and active harvester resistance to research and management, a devaluing of local harvesting traditions, and missed opportunities for collaboration. We reluctantly conclude that despite "New Forestry" co-management rhetoric, given existing institutional barriers and positivist scientific categories, NTFP workers will likely remain excluded from active roles in temperate forest research and management-contributing in turn to the ongoing legitimacy crisis of public and private forest management entities.

KEYWORDS. Chanterelle mushroom, forest management, legitimacy crisis, local knowledge, non-timber forest products

THE EMERGENCE OF NTFP HARVESTING AS AN "ISSUE"

"Whose woods these are I think I know. His house is in the village though . . ." ponders the poet. Across North America, particularly in Cascadia (the Pacific Northwest of British Columbia, Washington, Oregon, and northern California), unprecedented debate is raging about who uses and who should use these woods "lovely, dark and deep." To the surprise of many, harvesters of non-timber forest products (moss, salal, ferns, mushrooms, medicinal plants, etc., described in companion articles in this special issue and elsewhere) have emerged quite unexpectedly from these woods as stakeholders with substantial experience with and customary use-rights to vast tracts of public and private forestland.

In this paper we are concerned with the dynamics underlying and informing this historical moment, in which gathering of NTFPs in at least two regions of North America has suddenly emerged as an "issue." More precisely, why has this issue emerged now and over a specific concern with commercial harvesting? This has prompted the organizing of NTFP business associations, efforts to organize NTFP ("forest floor") workers, some agency and interagency reorganization, and stepped-up efforts to monitor and control harvesting. It has also consolidated a fledgling interdisciplinary research community, burrowing into various professional conferences and conducting research like that reported on in this special issue.

Drawing from a three year study of chanterelle mushroom harvesters on the Olympic Peninsula Biosphere Reserve (Washington, USA),[1] we identify and discuss, in no particular order, eleven factors–both background and emerging–which intersected in the late 1980s to bring NTFP harvesting to prominence as an issue in North America. In the second part of this paper we focus on one of these factors–eleventh in our list below–as central to the emergence of NTFP harvesting as an "issue" worthy of monitoring and control.

Absent NTFP harvesters in North America's historical narrative. Food foraging (reliance on extracting "wild" resources from "natural" environments) has been a prominent feature in the household economies of tribal and peasant peoples around the world at least since our emergence as a species some 100,000 years ago. Though Native Americans had long occupied the continent and relied on its resources, the notion of a virgin land developed by immigrant Europeans into an agricultural and industrial cornucopia is central to the "official" (late nineteenth century) version of the North American story. In contrast with the Spanish occupation of lands to the south, and despite the technology and knowledge transfer symbolized by Squanto and the Pilgrims, cultural and genetic mixing with the original inhabitants of what became Anglo America was minimal and minimized. State-sponsored Indian removal to open lands for Euro-American settlement started with the expansion of slave-based plantation cotton cultivation in the southeast and with the ever westward expansion of 18th and 19th century America. Though chains of knowledge were disrupted, indeed, usually decimated, Native American NTFP harvesting persisted.[2] Today, despite the preponderance of documentation about gathering traditions and often clearly stated harvesting rights in treaties, they receive scant legal recognition or scientific concern outside of tribes themselves. Active reliance on wild resources and Native

American knowledge was also critical to pioneer settlers over their 250-year westward migration to the Pacific. Many immigrants, whether uprooted European farmers or enslaved West Africans, brought cultural traditions of wild plant use with them and developed others while adjusting to life in the New World.[3] However, these did not translate prominently into the dominant cultural narrative and practices of present-day North America, a process Wendell Berry has termed the "unsettling of America" (1977).

Recent immigrants to North America. In contrast to this older immigration story, traditions and practices brought by more recent migrants have fed directly into the recent emergence of NTFP harvesting. The immigration and resettlement of various Southeast Asian peoples in the wake of the Vietnam War was focused on the West Coast, where vast public forestlands lay nearby and opportunities for picking are (were) unhindered by regulation. Though many stated they came from urban backgrounds, Cambodians, Vietnamese, Laotians and others began harvesting bear grass, floral greens and mushrooms in Washington and Oregon, followed more recently and increasingly displaced by Latinos, chiefly Mexicans (vid. Hansis 1998). Cross-cultural communication issues have amplified the struggles for managers to monitor and regulate harvesting in the Pacific Northwest. The role of Anglo mushroom buyers in connecting various Southeast Asians with commercial mushrooming has yet to be investigated. Though involving a mix of subsistence, recreational and commercial goals, the entry of recent immigrants has been more clearly tied to commercial harvesting and growing expansion of overseas markets.

Overseas markets and domestic demand. With a rapidly globalizing world economy, NTFP harvesting (especially in the Pacific NW) is increasingly significant in larger commodity circuits. Examined across the continent, however, it is clear that NTFP extraction has been part of local and regional commodity circuits for a much longer time, occasionally entering world markets. While much of the research and wider public focus in the Pacific Northwest is on the relatively new commerce in mushrooms, exporting of floral greenery is also very important and dates back to the 1930s on the Olympic Peninsula. In the east, the ginseng trade with Asia goes back over 275 years (Persons 1994) and the earliest woodland settlers surely practiced maple sugaring. Aging but increasingly affluent baby-boomers are demanding more natural means of averting the inevitable consequences of aging, leading to market niches opening for gourmet and organic foods and specialty products, including some NTFPs. More research is needed on the causes and sequence of the emergence of commercial harvesting for various types of products.

Growing awareness of extractive activity in many other parts of the world. In connection with growing environmental awareness in North America (see next), people became more aware of "sustainable" extractive activity in many other parts of the world, especially among Amazonian, Scandinavian, Slavic, Chinese, Japanese and various Southeast Asian peoples. The 1986 Chernobyl nuclear disaster affected large areas of prime mushrooming territory in the Ukraine, Russia, eastern Europe and Scandinavia. It became clear to many North Americans that this devastated NTFP harvesting, especially of mushrooms, in these areas in subsequent years. However, it is unclear the extent to which this stimulated market demand for North American chanterelles and other mushroom species (fide S. Alexander).

Ambivalent environmentalist sentiment about commodity extractive activity. The economic emergence of NTFP harvesting coincided with wider environmentalist sentiment against commodity extractive activity, particularly on public (forest) lands in the US Pacific Northwest. Taking the old-growth debate national brought to prominence the plight of highly visible endangered species like Northern Spotted Owl, and mobilized increasing concern over, if not outright opposition to, commercial NTFP harvesting. In our Olympic Peninsula study (Love, Jones & Liegel 1998), we found that amateur mycologists played an especially important role in this debate. North American environmentalists also adroitly linked these concerns with environmentally damaging development projects in the Amazon funded by international lending agencies, often with US taxpayer support. Violent suppression of rubber tappers and Brazil Nut collectors trying to protect a "sustainable" way of life led to the emergence in the late 1980s of Chico Mendes and tropical forests as key symbols of a global effort to protect forests. One result was ambivalently linked to commodity extraction of NTFPs in the establishment of extractive reserves in Rondonia, Acre and other parts of Amazonia.

Decline of timber harvest, especially in the Pacific Northwest. In accounting for the sudden emergence of NTFP harvesting, many observers would point to the steep decline in logging on western public forestlands, and concerns about old-growth dependent species in the most productive west side forests of the Pacific Northwest. Suddenly, unemployed loggers and mill workers began to turn to NTFP harvesting as a way to make ends meet and stay on the land in declining rural communities. While this is clearly an important proximate cause of the NTFP "issue," we are reluctant to place too much emphasis on the economic decline of timber-dependent communities. On the Olympic Peninsula we found that NTFP harvesting was very often part of the way of life of families in logging country. Families on the northwest side of the Peninsula had been harvesting mushrooms, berries and other NTFP products for generations, similar to the picture painted by Emery for the Upper Peninsula of Michigan (1998). Women, children and retired men all participated. As we note next, it was not an activity very often made public. Nevertheless, it was very much part and parcel of living in the woods, providing not only food and some cash, but also symbolizing a rural way of life and part of gift-giving among kin and friends. While NTFP harvesting may have intensified with declining employment in logging and mills (though we are unaware that this has been adequately documented), its sudden recognition by researchers, managers and others cannot be accounted for solely by the decline of the timber harvest, whether due to environmental restrictions or (more often) to mechanization.

Low occupational prestige of NTFP harvesting. Given its association with backwoods poverty and low-paid manual labor, NTFP harvesting has enjoyed low occupational prestige. On the Olympic Peninsula we found that though NTFP harvesters generally avoided open public identity, this masked an almost tribal sense of loyalty and pride–especially among mushroom pickers.

Geographic remoteness of most harvesting activity. Most NTFP harvesting was (and is) out of general public view, given vast forested landscapes and generally seasonal harvesting. Harvesters, especially commercial ones, often must travel great distances to find sufficient quantities of the products they seek. Consequently, we found that costs of keeping vehicles in good working condition figure prominently for many harvesters.

Harvesters' culture of secrecy. Commercial (and recreational) harvesters have a general desire and economic incentive to keep secret their search strategies and harvest areas. This is especially true of mushrooms, whose fruiting is episodic and much less predictable than that of most other NTFP species.

General mycophobia of Anglo culture. Especially with mushrooming, cultures like the US and Canada deriving from the British Isles have a general distaste for mushrooms, dismissing them all as undistinguishable "toadstools" (Arora 1990). This stands in sharp contrast with cultural attitudes of most other peoples of the world, a subject which has apparently received no scholarly attention.

Growing anti-science sentiment. Increased questioning of the authority of conventional modernist science in general North Atlantic culture has created a temporary space for the emergence of normally marginalized local knowledge and practices. Despite our primarily ethnographic leanings and misgivings about much of what post-modernists have actually achieved theoretically (vid. Callinicos 1990), we spend the remainder of this paper focusing on this last factor.

THE END OF PROGRESS AND POSITIVE KNOWLEDGE?

In trying to make sense of the way NTFP harvesting emerged quite suddenly as an issue in the early 1990s, we must self-critically turn the "scientific gaze" back on the research and policy communities to ask why it is news that many people in the core of the modern world system, such as in the Pacific Northwest and the Upper Peninsula of Michigan, make a living by harvesting forest resources other than timber? Why do the research, policy and management communities know so little about the people doing the actual harvesting work? Why is it only recently that the magnitude of revenues generated by NTFP harvesting are becoming appreciated and therefore worthy of managerial and scientific concern (e.g., Schlosser & Blatner 1995)? While this research blind-spot is certainly connected to the general rendering of humans from nature in the modern condition, it is more specifically connected to an academic division of labor in which the study of peoples engaged in non-wage work is marginalized and seen to be confined to the "developing, preindustrial, premodern world," in favor of a focus on waged labor in capitalized, corporately-organized industry. In the world view which this division reflects, one would not expect such quaint, uncivilized activity as NTFP harvesting in post-industrial "developed" societies.

All this is really the flip side of the realization that one use of our forestlands–large-scale clearcut logging–assumed such primacy in temperate forestry that other ways to use forest resources were largely ignored if not excluded. The everyday practices of people as diverse as forestry professionals in and out of government, business and academia, NTFP harvesters, buyers and consumers, labor organizers, and others are affected by the period of time we are in–an historical juncture between convincing models of how forestlands, both public and private, should best be managed (Hirt 1994; Franklin 1993; vid. Walker & Daniels 1996). This is not just an institutional crisis in which public and private forest-managing institutions find themselves, but also part of a larger cultural crisis of post-Cold War North Atlantic civilization.

The modernist consensus, expressed in forestry in the multiple use paradigm that until recently dominated research, management, and policy, reached its zenith during the Cold War struggle between superpowers. It privileged a technocratic scientific discourse of rational experts and spawned an entire university research establishment tied in many ways to it. It came at the expense or exclusion of alternative or local knowledge, not only internationally but also within our own country. Minority understandings of situations, their values, their version of events, their narratives, weren't necessarily destroyed, but rather subordinated to a larger, hegemonic discourse privileging scientific ways of knowing.

> [But] over the last thirty years, what might be called the master narrative of the national forests has changed. In the original story there was once a vast and bountiful nature. Americans exploited this bounty to build a civilization, but abundance bred waste and carelessness. Far-sighted men, recognizing that the resources were not unlimited, wisely saved a remnant of the original abundance by withdrawing it from the public domain. Carefully nurtured, these lands have yielded profusely as skilled managers have made sure that what is taken is replenished. [. . .] recent studies of the national forests have not been kind to this narrative. (White 1992:173)

Given the relatively uncontrolled resource destruction occurring around the turn of the last century, there were good reasons why the consensus was constructed and lead to chartering the public forestlands system. This earlier "custodial" paradigm shifted in the post-World War II period to the multiple use paradigm, involving a "conspiracy of optimism" in which a multiple-use rhetoric masked a dominant logging use of both private and public forestlands (Hirt op cit.). In neither of these "old forestry" frameworks was there a place for or concern with non-timber forest products harvesters. As noted above, Native American uses of non-timber resources were marginalized and ignored, if not terminated. While commercial harvesting of floral greens or mushrooms had begun by the 1920s and developed some during the Depression, it was barely perceived by forest managers concerned with "getting the cut out." NTFP harvesters have never been part of the imagined community of forest managers, though District Rangers and others close to the ground realized and sometimes even managed for these local uses.

By the 1980s the volley of new claims on public forestlands converged into a growing legitimacy crisis for institutions operating under the multiple use paradigm. This became a "legitimacy" crisis because a critical mass of the public came by the late 1980s to suspect that multiple use rhetoric in fact disguised management practices and policies favoring large scale clearcut logging and fiber production. Formerly acceptable practices now became unpopular in the shifting lens of popular culture. This legitimacy crisis came to be felt by academics in the forest research community, most of whom remained (and remain) wedded to scientific positivism and the multiple use forestry paradigm (vid. McEvoy 1992). One of the consequences of the conspiracy of optimism (Hirt op cit.) was not only a backlash against agency (and industry) promotion of excessive timber cutting, but also a delegitimizing of science and deep questioning of the privileged position of technical experts. Other, subordinated knowledge bubbled up to public awareness.

The current legitimacy crisis of public agencies (USFS, BLM, etc.) and the parallel suspicion of corporate timber practices, other timberland managers and academic researchers are, in turn, located in a broader paradigmatic or discursive shift. We seem to be between convincing narratives about how the world works, variously conceived as "paradigms" (Kuhn 1972), "plausibility structures" (Berger and Luckmann 1966) or "guiding fictions" (Shumway 1991). With the collapse of the Soviet Union and the end of the Cold War, North Atlantic society (North America and western Europe) appears to be at a pivot point in our civilizational history, in which even the possibility of progress is increasingly questioned:

> The idea of progress [. . .] began with classical Greece and its fascination with knowledge, a fascination that was appropriated and put to intellectual and practical use by Christianity. From the early church fathers through the high Middle Ages and into the Puritan seventeenth century of Isaac Newton and Robert Boyle, there was a confidence that ever-expanding knowledge held the promise of something like a golden age. Though often in militantly secular form, this confidence drove also the Enlightenment, which was living off the capital of Christian faith in historical purpose. The assumed link between knowledge and progress explains [. . .] the liberal belief in'education' as the panacea for human problems, paving the road to utopia. But by the 1970s [. . .] all the talk was about the limits of knowledge, the end of scientific inquiry, the unreliability of claims to objective truth. The curtain was falling on the long-running show of modernity and progress. What would come to be called 'postmodernity' was waiting in the wings. [. . .] For many centuries, the argument was that knowledge equals progress, and now–or at least many were saying–advances in real knowledge were coming to an end. In 1978 an entire issue of the journal *Daedalus* was devoted to articles by scientists on 'The Limits of Scientific Inquiry.' Not only does science no longer have the cultural and even moral authority that it once enjoyed, the contributors noted, but many scientists are filled with doubts about their own enterprise. Some went so far as to suggest that we may be witnessing a reversal of roles between science and religion, with the ascendancy of the latter in providing a stable definition of our historical circumstance. (Neuhaus 1999:22-23)

A deep sense that "something is changing" pervades most aspects of our lives. Old left-right oppositions and resultant coalitions are giving way to new oppositions, such as between "stasists" (from reactionaries who resist change to technocrats who want to control change) and "dynamists" (those who encourage innovation and relish change) (Postrel 1998). When Pat Buchanan and Jeremy Rifkin start agreeing on globalization, you know something is changing.

The legitimacy crisis of forest management institutions and the putative larger civilizational break in which this is all located are in turn connected with a global system of unprecedented scale and integration. The modern world system is increasingly characterized by a higher degree of economic than political centralization, reliance on wage labor, finance capital seeking maximum returns on investment, relatively unfettered private accumulation and the establishment of political apparati to facilitate that accumulation.[4]

How ironic, then, that "hidden" NTFP harvesters are forest workers more closely tied to global markets than many other sectors in the Pacific Northwest economy, given how significantly linked the wild mushroom industry is to export markets in Japan and western Europe (Schlosser and Blatner op cit.). NTFP harvesting cannot be understood apart from the global markets and forces with which it is intimately connected; as old occupational niches are shrinking, new spaces are being created in which some people are finding employment and subsistence, however precarious it might be. But there is both downward and upward socio-economic mobility in this changing sector. Some harvesters are economically and socially marginalized people thrown down onto a subsistence base, in whole or in part, while others are using harvesting as an interim step up to more secure and stable occupations.

Economically or socially marginalized people frequently experience dislocation under the dynamic, if not volatile conditions of the increasingly stratified modern world system. In the swelling cities as well as the countryside of the unevenly developing periphery of this world system, urbanization without commensurate industrialization is creating havoc. In the "developing world" those remaining in rural areas, such as peasants with insecure resource tenure often fall back on a subsistence resource base, if they can. Migrants streaming northward from Mexico and Central America seeking economic security (Chavez 1998) as NTFP harvesters (or in other sectors) in the Pacific Northwest and local pressure on resources in such places as Chiapas are twin aspects of this same socio-economic upheaval.[5]

However, it is not only in the periphery of the modern world system–places like Chiapas or Rondonia–where such dislocations are taking place, but also in the internally-colonized interstices of the core of this same global system (Nash 1994)–such as the Pacific Northwest. In the current economic restructuring and capital flight offshore, factories close or move and employment opportunities become increasingly precarious and volatile. Reorganization and automation in the timber industry have had profound effects, especially on small mill-dependent towns, exacerbated by government policies and automation sharply reducing timber harvest rates throughout the region (Carroll 1995:1-8; Brown 1995). How these changes are connected with NTFP harvesting and access to subsistence resources are aspects that have only begun to be investigated systematically.

IMPLICATIONS FOR MANAGEMENT AND POLICY

As we have begun to see, NTFP harvesters did not figure in the "old forestry" models of the 19th century: neither in the subsequent custodial paradigm under which the system of state-managed forest reserves was chartered, nor in the post-WWII timber harvest-oriented multiple use paradigm. By the late 1980s the multiple use model was falling apart. A "new forestry" ecosystem management paradigm is now under construction. But it is unclear how, or even whether, NTFP harvesters will achieve recognition as legitimate stakeholders at the table this time around. McLain and Jones (2001) discuss how harvesters, buyers, forestry professionals, labor organizers, academics, and others can engage the institutional and discursive openings temporarily present.

Despite the fact that both floral greens collecting and wild edible mushrooming are multimillion dollar industries and employ thousands of people in the Pacific Northwest (ibid.), NTFP workers' practices or knowledge are scarcely being incorporated into management practices or policy on public or private forestlands. For example, various authors in a recent discussion about managing forest ecosystems to conserve fungal diversity complain of shortages of funding and trained fieldworkers, yet few raise the possibility, let alone the benefits of turning for help to knowledgeable commercial harvesters, who typically remain the "other" outside ordinary scientific practice (Pilz & Molina 1996)[6] (however, vid. Emery, 2001).

Though there is growing recognition of the economic significance of non-timber resources (vid. Alexander & McLain, 2001), our experience on the Olympic Peninsula and related projects leads us to remain skeptical that NTFP harvesters' concerns and perspectives will find much place in the coming paradigm. For example in the case of wild edible mushrooming in Washington, to date better organized recreational pickers have had more success than commercial harvesters in affecting policy (vid. McLain, Shannon and Christensen 1998): a ban on commercial mushroom harvesting on state lands was instituted in 1992 (though it expired in 1997) and more restraints are being actively contemplated. Also, severe cutbacks in agency budgets appear to be leading to a retreat to known and efficient management practices; acknowledging the importance of NTFP resources would require management agencies to monitor them biologically–a costly undertaking. How much more would be demanded of agencies systematically to acknowledge harvesters' understandings, let alone the challenges that would be posed in incorporating harvesters as co-managers?[7]

In this discursive break, the technical knowledge of wild mushroomers need not be overstated, but does need pointing out. Harvesters represent a large and diverse contingent of people working in the forest making daily observations, often in semi-systematic form, driven by the need to understand how to avoid hindering or ruining the resource from which they make their living. This suggests an innate component of sustainable economic experimentation is taking place with many harvesters. Without romanticizing, we are coming to understand that some pickers and buyers–those most engaged in NTFP activity–are in many ways ahead of the research community in understanding the biology and ecology of mushrooms, not to mention the organization of picking, grading, and marketing them.[8]

It is becoming obvious in our fieldwork that most harvesters are not just out in the forest blindly extracting the resource without regard for biological and economic consequences. For example, newly emerging mushrooms are often left by harvesters to mature into larger mushrooms. Harvesters indicate this is because it is usually not economically worthwhile to harvest small mushrooms, and that small mushrooms may have to be a certain size (or form in some cases) to disperse their spores for reproduction. Though harvesters could get more weight, and thus more money by pulling mushrooms, most chanterelle harvesters insist that cutting does less damage to the mycelium and further claim that chanterelles will sometimes regenerate from a cut stalk. Some harvesters will go as far as to leave patches lie fallow under the assumption that this will result in greater productivity in the future. It isn't uncommon to hear a harvester speculating on the importance of leaving mushrooms for other elements of the ecosystem (e.g., wildlife, insects, possible floral interdependence). We have observed other harvester experiments at increasing productivity, including stomping rotten mushrooms in the patch and returning trimmings and other mushroom wastes back to the forest to encourage spore dispersal and nutrient recycling. A few harvesters have said they tried to inoculate forest areas with isolated spores in hopes of "seeding the ground."

In sum, commercial harvesters are fulfilling at least two important roles that could be an immediate basis for a relationship with scientists and land managers. First, all harvesters regularly go into the woods off-trail to collect products (in this case, chanterelle mushrooms), all the while casually monitoring habitat for changes and patterns related to the non-timber forest product they are collecting. Given their frequent, often almost daily sampling and monitoring during the harvest season, commercial harvesters' ethnomycological knowledge appears to be stronger in the ecology and phenology of species they regularly harvest, but much weaker in other aspects of species biology or taxonomy. Commercial harvesters may also be able to assist mycological researchers in finding rare species of mushrooms, though we did not attempt to plumb this aspect of harvester knowledge. Second, as described above, some harvesters are consciously investigating processes that increase resource productivity. Harvesters are not as constrained by the predominantly analytic characteristic of much scientific research on forest ecosystems, which isolates specific, individual parts to be monitored and regulated from complex ecosystem processes. Their knowledge and experimental strategies would be useful in providing a more complete understanding of NTFPs.

AUTHOR NOTE

Thomas Love's economic and ecological anthropological fieldwork in the central Andes has resulted in many conference papers, several articles, a book and a forthcoming monograph on the social construction of regional identity in southern Peru. His fieldwork on forest sustainability issues has revolved principally around non-timber forest issues in NW North America and SE Peru.

Eric T. Jones' primary area of research is cultural dimensions of non-timber forest product harvesting in temperate forests in the U.S. Pacific Northwest and abroad.

The authors thank many colleagues with whom they discussed these issues, some of whom commented on earlier drafts of this paper, among them Sue Alexander, Kristin Barker, Alfred Darnell, Dick Hansis, Leon Liegel, Joel Marrant, Rebecca McLain, Randy Molina, Carol

Mortland, Jeri Peck, Jeff Peterson, Dave Pilz, Ron Post, Ellen Short, George Stankey, Nan Vance, Bettina von Hagen and Alex Wegmann.

NOTES

1. The MAB Chanterelle Mushroom Project was a three-year (1994-1996) interdisciplinary research project funded by the Man and the Biosphere Program, U.S. Department of State ("Biological, Socio-Economic and Managerial Concerns of Harvesting Edible Mushrooms on the Olympic Peninsula and the Southern Appalachians"). In a companion piece to this article (Love, Jones & Liegel 1998) we describe more fully and empirically the social and economic organization of this harvesting, including data gathering methods used. Standard research protocols regarding confidentiality and privacy were followed scrupulously.

2. Research is needed on the importance of NTFPs as stress buffers during these times.

3. Important folk traditions related to NTFP harvesting were developed by slaves and persist in the southeast today (fide C. Anthony). White Appalachian folk traditions center on use of local plants and animals of the hills and hollows. More work on the social and cultural history of NTFP harvesting remains to be done by environmental historians and others, particularly contrasting the longer acculturation experience of the Eastern Woodlands Indians with that of Native populations to the west.

4. This is hardly the place to embark on a social history of the modern world system and the place of scientific ways of knowing in it. Wolf (1982) and Wallerstein (1979) are good places to begin and to find further references.

5. For example, in highland Chiapas, Mexico, indigenous groups have been intensively studied by anthropologists as if they were repositories of some pristine pre-Columbian cultural essence. Recently investigators (e.g., Wasserstrom 1989; Collier & Quaratiello 1994) have demonstrated how growing peasant pressure on their environment, for example in deforestation and emigration to the rainforest frontier, cannot be understood apart from the pressures from the larger political economic dynamics which condition peasant livelihoods.

6. Commercial harvesters are not described in neutral or mildly positive terms until pp. 79 and 92 in Pilz and Molina (1996). Only on p. 99 are commercial harvesters acknowledged for their skill and the contributions they could make to management and research.

7. The Adaptive Management Area (AMA) concept of the President's Northwest Forest Plan is one ambitious attempt to move toward co-management, though rhetoric appears to be far outpacing reality. Love is currently investigating this topic in the North Coast Range AMA in western Oregon.

8. It is important to point out in this context that mycological research in North America has typically been severely under-funded. This is certainly related to the general mycophobia of Anglo culture, so that mushrooming has not been recognized either as a worthwhile activity or an activity worth knowing about (vid. Arora 1990).

REFERENCES

Alexander, S. and R.J. McLain. 2001. An overview of non-timber forest products in the United States today. Journal of Sustainable Forestry 13(3/4): 59-66.

Arora, D. 1990. Mushrooms demystified. Ten Speed Press, Berkeley, CA.

Berger, P. and T. Luckmann. 1966. The social construction of reality: a treatise in the sociology of knowledge. Anchor Books, Garden City, NY.

Berry, W. 1977. The unsettling of America: culture and agriculture. Sierra Club Books, San Francisco, CA.

Brown, B. 1995. In timber country: working people's stories of environmental conflict and urban flight. Temple Univ. Press, Philadelphia, PA.

Callinicos, A. 1990. Against postmodernism: a Marxist critique. St. Martin's Press, New York, NY.

Carroll, M. 1995. Community and the Northwest logger: continuities and changes in the era of the spotted owl. Westview Press, Boulder, CO.

Chavez, L. 1998. Shadowed lives: undocumented immigrants in American society, 2nd ed. Harcourt Brace College, Fort Worth, TX.

Collier, G.A. & E.L. Quaratiello. 1994. Basta! Land and the Zapatista rebellion in Chiapas. The Institute for Food and Development Policy, Oakland, CA.

Emery, M.R. 1998. Invisible livelihoods: non-timber forest products in Michigan's Upper Peninsula. UMI Dissertation Services, Ann Arbor, MI.

Emery, M.R. 2001. Who knows? Local non-timber forest product knowledge and stewardship practices in northern Michigan. Journal of Sustainable Forestry 13(3/4): 123-139.

Franklin, J. 1993. The fundamentals of ecosystem management with applications to the Pacific Northwest. *In:* Aplet, G.H., N. Johnson, J.T. Olson & V. Alaric (eds.). Defining sustainable forestry. Island Press, Washington, DC.

Hansis, R. 1998. A political ecology of picking: non-timber forest products in the Pacific Northwest. Human Ecology 26(1): 49-68.

Hirt, P. 1994. Conspiracy of optimism: management of the national forests since World War II. Univ. of Nebraska Press, Lincoln.

Kuhn, T. 1972. The structure of scientific revolutions, 2nd ed. Univ. of Chicago Press, Chicago, IL.

Love, T., E. Jones, and L. Liegel. 1998. Valuing the temperate rainforest: wild mushrooming on the Olympic Peninsula Biosphere Reserve. AMBIO–A Journal of the Human Environment, Special Rpt. #9, pp. 16-25.

McEvoy, A.F. 1992. Science, culture, and politics in U.S. natural resources management. Journal of the History of Biology 25(3): 469-486.

McLain, R.J. and E.T. Jones. 2001. Expanding non-timber forest product harvester/buyer participation in Pacific Northwest forest policy. Journal of Sustainable Forestry 13(3/4): 147-161.

McLain, R., H.C. Christensen, and M.A. Shannon. 1998. When amateurs are the experts: amateur mycologist and wild mushroom politics in the Pacific Northwest, USA. Society & Natural Resources 11 (1998): 615-626.

Nash, J. 1994. Global integration and subsistence insecurity. American Anthropologist 96(1):7-30.

Neuhaus, R.J. 1999. The idea of moral progress. First Things 95:21-27.

Persons, W.S. 1994. American ginseng: green gold. Bright Mountain Books, Inc., Asheville, NC.

Pilz, D. and R. Molina, eds. 1996. Managing forest ecosystems to conserve fungus diversity and sustain wild mushroom harvests. Gen. Tech. Rep. 371, U.S. Department of Agriculture, Forest Service, Pacific Northwest Research Station, Portland, OR. 104 pp.

Postrel, V. 1998. The future and its enemies: the growing conflict over creativity, enterprise, and progress. Free Press, New York, NY.

Schlosser, W. and K. Blatner. 1995. The wild edible mushroom industry of Washington, Oregon and Idaho: a 1992 survey of processors. Journal of Forestry 93(3): 31-36.

Shumway, N. 1991. The invention of Argentina. Univ. of California Press, Berkeley.

Walker, G.B. & S.E. Daniels. 1996. The Clinton administration, the Northwest forest conference, and managing conflict: when talk and structure collide. Society and Natural Resources 9:77-91.

Wallerstein, I. 1979. The capitalist world-economy. Cambridge Univ. Press, Cambridge, MA.

Wasserstrom, R. 1989. Rural labor and income distribution in central Chiapas. *In:* pp. 101-117. Orlove, B.S., M. Foley & T.F. Love (eds.). State, capital and rural society: anthropological perspectives on political economy in Mexico and the Andes. Westview Press. Boulder, CO.

White, R. 1992. Indian land use and the national forests. *In:* pp. 173-179. Steen, Harold K. (ed.). Origins of the national forests: a centennial symposium. Forest History Soc., Durham, NC.

Wolf, E. 1982. Europe and the people without history. Univ. of California Press, Berkeley.

Who Knows? Local Non-Timber Forest Product Knowledge and Stewardship Practices in Northern Michigan

Marla R. Emery

SUMMARY. Non-timber forest product (NTFP) literature frequently laments the absence of an information base for policy and management decisions. While formal scientific data on the biological and social ecologies of most NTFPs are limited to nonexistent, long-time gatherers often have extensive experiential knowledge bases. Researchers and managers may overlook this expertise because of assumptions about the nature of knowledge and the identity of individuals who possess valuable information. These assumptions are explored and contrasted to the concept of local knowledge. A case study of gatherers in Michigan's

---

Marla R. Emery is Research Geographer, USDA Forest Service, Northeastern Research Station, 705 Spear Street, P.O. Box 968, Burlington, VT 05402-0968 USA (E-mail: memery/ne_bu@fs.fed.us).

The author wishes to thank A. L. (Tom) Hammett and Alan Pierce for their valuable comments on the paper. The author is also indebted to Beth Lynch, former botanist with the Great Lakes Indian Fish and Wildlife Commission, Dana Richter, Research Scientist at Michigan Technological University's School of Forestry and Wood Products, and Jan Schultz, Forest Plant Ecologist with the Hiawatha National Forest, for their assistance in identifying the species and scientific names of Upper Peninsula NTFPs. The research on which this paper is based was generously supported by the USDA Forest Service's Northern Global Change Program and Hiawatha National Forest.

[Haworth co-indexing entry note]: "Who Knows? Local Non-Timber Forest Product Knowledge and Stewardship Practices in Northern Michigan." Marla R. Emery. Co-published simultaneously in *Journal of Sustainable Forestry* (Food Products Press, an imprint of The Haworth Press, Inc.) Vol. 13, No. 3/4, 2001, pp. 123-139; and: *Non-Timber Forest Products: Medicinal Herbs, Fungi, Edible Fruits and Nuts, and Other Natural Products from the Forest* (ed: Marla R. Emery, and Rebecca J. McLain) Food Products Press, an imprint of The Haworth Press, Inc., 2001, pp. 123-139. Single or multiple copies of this article are available for a fee from The Haworth Document Delivery Service [1-800-342-9678, 9:00 a.m. - 5:00 p.m. (EST). E-mail address: getinfo@haworthpressinc.com].

Upper Peninsula found that many possess extensive knowledge of the products they harvest and observe stewardship practices to assure their sustained availability. The paper is illustrated by descriptions of four gatherers and concludes with recommendations for incorporating the local knowledges of individuals from a variety of cultures into policy, research, and management. *[Article copies available for a fee from The Haworth Document Delivery Service: 1-800-342-9678. E-mail address: <getinfo@haworthpressinc.com> Website: <http://www.HaworthPress.com>]*

KEYWORDS. Non-timber forest products, local knowledge, sustainable use, Michigan

> The major problem related to research concerning NTFP is the limited availability of information related to sustainable utilization . . . The difficulty stems from a singular lack of hard scientific data . . . (Tewari and Campbell 1995, p. 57)

INTRODUCTION

Concerns about the (un)sustainable harvest of non-timber forest products (NTFPs) lurk in virtually every discussion of their potential as a tool for preserving biodiversity or promoting economic development. Laments about the lack of information on which to base sustainable management are nearly universal in the expanding body of books and articles on NTFPs. These concerns are founded in well-documented cases of unsustainable NTFP use, especially following the development of a commercial market for products (Homma 1996). Yet the work of ethnobotanists and anthropologists documents extensive knowledge of NTFPs and their long-term harvest by gatherers in the developing world and among indigenous peoples in North America. For example, Grenand and Grenand (1996) and Kainer and Duryea (1992) have described long-term reliance on forest-based goods and social processes that develop and maintain knowledge of NTFP use, ecology, and management in tropical South America. Publications by Anderson (1999), Gottesfeld (1994), Nabhan et al. (1991), Richards (1997), and Turner (1998) document the ecological knowledge and plant technologies of Native American and First Nations peoples in North America.

Why, then, do researchers and resource managers feel that so little is known about what is probably the oldest interaction between humans and forests–the gathering of wild plant material? It is true that as a group, researchers and managers possess only rudimentary knowledge of the biological and social ecologies of NTFPs. However, as the ethnobotanical and anthropological literature suggests, we ignore an extensive body of knowledge when we maintain that information does not exist. Examination of assumptions about the nature and location of knowledge may shed light on this disconnect and address the need for information on which to base strategies for sustainable NTFP use.

Using a case study from northern Michigan's Upper Peninsula (UP), this paper suggests that considerable information about NTFPs currently exists in the United States among gatherers of diverse ethnic backgrounds. A brief discussion of theoretical work on knowledge and its implications for NTFP research serves as prelude to a description of results from fieldwork that revealed local knowledge and stewardship norms among gatherers in the UP. These are illustrated by four examples of gatherer knowledge and stewardship practices. The paper concludes with recommendations for the inclusion of local knowledge in NTFP research, planning, and management.

KNOWLEDGES AND KNOWERS

Scholars of the sociology of science and feminist analysts of the history of knowledge and power have examined the rise of science in the history of human efforts to understand and act within the material world. They note that the scientific drive to discover and describe patterns in nature and to use that information to influence both social and biophysical processes is not historically unique. However, science is distinguished from other systems of knowledge by its commitment to universal principles or laws, their discovery through examination and manipulation of the individual parts of the object under study, and subsequent broad application across time and space. These scholars also chronicle the processes by which science has supplanted other knowledge systems and scientists have attained the status of unique authority (e.g., Haraway 1989; Harding 1986; Merton 1973; Mulkay 1979).

Such analyses have coincided with recognition of the often unanticipated and undesirable consequences of the science-based Green Revolution and development projects such as massive dams and flood control programs. A growing body of literature draws on these theoretical insights and empirical case studies to suggest the existence of local knowledge and the need for its incorporation into the information base for sustainable policy, planning, and management (DeWalt 1994; Flora 1992; Kimmerer 2000; Kloppenburg 1991; Murdoch and Clark 1994).

Within this context, local knowledge is defined as information on the social and ecological characteristics of a particular location, which is acquired through experience (particularly physical labor) in a place over time. It is produced through extended observation, direct interaction, and, sometimes, *in situ* experimentation. Local knowledge is characterized by rich, specific detail on the patterns and variations in the aspects of a place with which the knower has direct contact. The highly personalized nature of this knowledge and its boundaries in space and time may be expanded by the inter-generational transfer of local knowledge. However, its fundamental characteristic is that it is produced and reproduced through direct experience (DeWalt 1994; Kloppenburg 1991; Murdoch and Clark 1994). Consequently, there is no singular local knowledge. Rather, more accurately there are local knowledges. To the extent that the activities of individuals are conditioned by social characteristics such as age, gender, class, and ethnicity, local knowledges may be similarly differentiated (Feldman and Welsh 1995; Mosse 1994).

Within the context of this ongoing theoretical discussion, it seems worthwhile to ask a number of questions related to sustainable NTFP use. Are there valuable sources of information on NTFPs outside of formal science? If other knowledge bases exist, should they be consulted in planning and managing NTFP programs? And, if local knowledges are potential sources of information, how might they be incorporated? Finally, who possesses valuable NTFP knowledge?

As noted in the introduction, a number of researchers have explored local NTFP knowledge and stewardship practices in the developing world and among indigenous peoples in the United States and Canada. The location of these case studies in distant places and historically marginalized ethnic groups may lead U.S. researchers and managers to assume that local NTFP knowledge is the exclusive domain of 'othered' cultures. That is, that knowledgeable gatherers must be people whose national and cultural identities are clearly different than those of most researchers and managers. However, any such assumptions must be made with caution. McLain and Jones (1997) and Love et al. (1998) have recently described the mushroom harvesting techniques and stewardship practices of Pacific Northwestern gatherers with a variety of ethnic backgrounds. And, as described below, a year of ethnographic research on the role of NTFPs in household livelihoods in Michigan's UP revealed extensive local knowledge of NTFPs among both Native American and European American gatherers (Emery 1998a).

NTFP USE IN MICHIGAN'S UPPER PENINSULA

Michigan's UP has the high forest area and low human population density sometimes associated with sustainable NTFP use. Its approximately 10.5-million-acre land base is 83.9% forested (Schmidt, et al. 1997) and has an average of fewer than 18 persons per square mile (U.S. Census Bureau 1990). Second growth maple-beech-birch and spruce-fir constitute the major forest types in the region (44.4% and 24.8% of timberlands, respectively; Schmidt et al. 1997). The ethnic composition of the human population includes a number of Anishinabe tribes (also known in English as Ojibwa or Chippewa) and European Americans. Many of the latter are descendants of loggers and miners who were recruited from Cornwall, Finland, Italy, Sweden, and French-speaking Canada during the late 1800s and early 1900s to work in the region's once-flourishing timber, iron, and copper industries (Cleland 1992; Dunbar 1965; Martin 1986).

Between August 1995 and July 1996, more than 400 hours of semi-structured interviews were conducted with gatherers, buyers, and public and private land managers in the UP to learn what NTFPs were harvested there and the role they play in gatherers' household livelihoods. Results reported here are based on information provided by 43 individuals about their personal gathering activities and experiences. Of these, 10 identified themselves as Native American and 33 as European American. Questions asked during the interviews focused on what the individual gathers, how each NTFP is used, ecological characteristics associated with products, harvesting techniques and norms, and how the gatherer learned these skills.

By the end of the year a list had been compiled of 139 NTFPs from over 100 botanical species that gatherers reported personally harvesting in the region's forests and associated open lands (Emery 1998a).[1] These products are used as medicine, food, craft or decorative materials, and ceremonial items. They provide for household needs through both nonmarket (personal consumption and gifts) and market means (sale in raw and processed forms). Nonmarket uses constitute nearly two-thirds (64%) of gatherers' NTFP livelihood strategies, with edibles the single largest functional use among these. Market uses account for the balance of livelihood uses, with the sale of processed craft or decorative items most frequently mentioned.

As in many natural resource dependent areas, unemployment is chronically high in the UP[2] and NTFPs are part of livelihood strategies by which households rely on contributions from a number of individuals and sources to meet their needs. NTFPs are particularly important when formal employment and income are inadequate to meet household needs and for individuals with limited access to jobs by virtue of age, gender, or disabilities of various types. Not surprisingly, NTFP livelihood strategies are usually displaced by formal employment opportunities. However, gathering also has important cultural and recreational dimensions and is pursued by individuals and households even when there is no pressing financial need (Emery 1998b). At such times, gathering allows people to enjoy time in the woods, keep their NTFP skills alive, and pass them on to younger friends and family members while deriving immediate practical benefit from the activity.

Unlike the Pacific Northwest, recent booms in the commercial market for many NTFPs have not reached the UP. Although some NTFP businesses in the region do sell products in the national and/or global markets, capital and ownership continue to be local, as does virtually all labor. Consequently the social structures and technologies of NTFP harvesting appear to have remained relatively stable over the last several decades.

KNOWLEDGE REQUIREMENTS

Whether for personal consumption or sale, gatherers must have substantial knowledge and skills to convert forest plant matter into livelihood resources. This knowledge may be thought of as having three dimensions–use, ecology, and economics (Figure 1). All three are required for an NTFP to make meaningful material and cultural contributions. The general characteristics of local NTFP knowledge are described below and illustrated in the four examples with which this section concludes.

FIGURE 1. UP gatherers' local NTFP knowledge

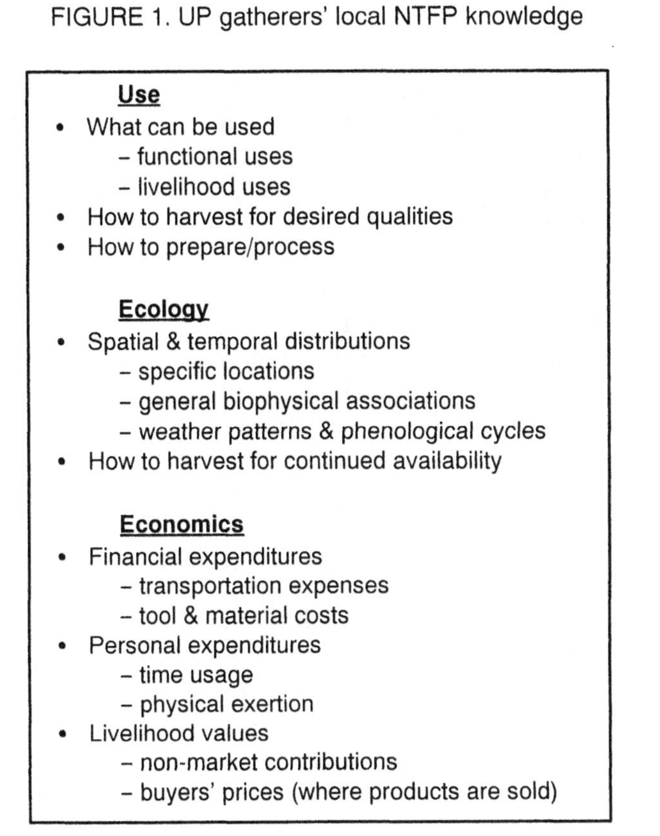

Use knowledge has three key features: what forest plant matter may be used and how, how to harvest a product for the desired qualities, and how to prepare or process it. For example, such knowledge is required to distinguish edible and poisonous plants, useful and useless or potentially dangerous medicinal doses, and tree barks that will make aesthetically pleasing and durable containers. When NTFPs are a source of cash income, gatherers must know which products have commercial value and who buys them. Regardless of their livelihood use (e.g., market or nonmarket), gatherers must also know how to harvest NTFPs so that they have the desired qualities and how to prepare them so that they are fit for their intended functional uses.

Ecological knowledge, the second dimension, is also critical because gatherers must know how to find a product in quantities sufficient to make harvesting and processing worthwhile. This knowledge has spatial and temporal dimensions. Spatial knowledge includes specific locations in which a product is known to occur as well as the general biophysical characteristics associated with it. The latter may include land-use history, soils, hydrology, aspect, elevation, and co-location with other plant and tree species.[3] Temporal knowledge is also essential because many NTFPs are ephemeral, with the plant present in the forest only briefly or possessing the desired property (such as nuts, fruits, or running sap) for a short period of time. Thus, gatherers must know *when* to look as well as *where* to look. This includes knowledge of the annual and interannual occurrence of products in association with weather and climate patterns and the co-occurrence and phenology of associated forest species, especially easily visible phases such as budding, leaf out, and flowering.

Detailed ecological knowledge is facilitated by small gathering ranges. Of the 21 individuals for whom data were collected on the distances they typically cover to harvest NTFPs, 12 (66%) gather within a 30-mile radius of their residence, with 75% of these (9 individuals) reporting that they stay within 10 miles. Thus, most harvesting occurs within the territory of gatherers' day-to-day lives, and they often have repeated experience with productive locations over time.

The final dimension, economic knowledge, is also critical if NTFPs are to provide livelihood resources. If the intent is to provide for personal consumption or gifts, expenditures of time, money, and personal energy must be reasonable in comparison to the material and cultural benefits derived from harvesting. When planning to sell an NTFP raw, the gatherer must know the prevailing buyers' prices and calculate the costs of transportation, tools, and labor to be sure the time and effort will provide needed income. Likewise, crafters and others who sell processed NTFPs must know the costs of materials, transportation, and gathering and production time as well as the prices they can charge for their goods. Many of the individuals interviewed were able to specify each of these costs and reported factoring them into decisions about when and where to expend their efforts.

STEWARDSHIP PRACTICES

When the intent is to rely on NTFPs on an ongoing basis, gatherers must know how to harvest products so that they will continue to be available year after year. This is critical to assure an adequate population of the desired species within a reasonable range. Everyone interviewed reportedly had relied on NTFPs for several years, if not decades, and hoped that both they and their grandchildren would be able to continue to do so. When asked, "Are there things you will and will not do when you pick "NTFP X"? and Do you have any rules for yourself?", many gatherers responded by describing stewardship practices that are intended to meet the dual goals of plant sustainability and gathering sustainability.

Specific stewardship practices vary with the biological characteristics of particular products and individual harvesters. However, most UP gatherers articulated one or more general norms or rules that they observe when harvesting NTFPs (Figure 2). Several people mentioned rotating their gathering from individual trees or in particular areas over periods of 3 to 10 years.[4] Other gatherers monitor plant biology as a key determinant of the gathering times for several products, looking for physiologic characteristics that indicate a product can be harvested without causing long-term damage or interfering with reproductive cycles. Nearly a dozen people stressed the importance of gathering selectively at nesting scales from individual plants to the landscape level. These gatherers will harvest only some leaves on a branch, some plants from a patch, and some patches in a landscape in any year. Minimizing harvest impacts was mentioned more frequently than any other norm. Gatherers talked about taking only what they will use or need and leaving no visible sign of their activities, indicating that they

FIGURE 2. UP gatherers' stewardship practices

- **Rotate gathering** over multiple years on same area and/or individual plants.
- **Time gathering** in accordance with plant biology.
- **Gather selectively** at nesting scales from individual plants to landscapes, never take everything.
- **Minimize harvest and impacts**, taking only what is needed/will be used and/or leave no damage or visible sign of gathering activity.
- **Maximize utilization**, creating and leaving no waste.
- **Promote growth** through harvest techniques or other intentional propagation.
- **Protect sites** from over-harvesting by closely guarding knowledge of their locations.

consider it unacceptable to damage soils and plant matter or leave waste of any kind. Closely related was an emphasis on maximizing utilization. A number of people stated that they endeavor to use everything they collect. A few gatherers consciously employ harvesting techniques that they believe promote further biomass growth or engage in seed dispersal and other propagation efforts. Finally, many people keep their sites a closely guarded secret in an effort to protect them from over-harvesting by others.

Examples 1-4 illustrate gatherers' knowledge and stewardship practices:

EXAMPLE 1. Knowledge–livelihood uses, economics of expenses and market values
Stewardship practices–promote growth

Allan[1]

Allan gathers five NTFPs, which he sells to a local microenterprise. Almost every year since he was 12 years old, he has cut balsam (*Abies balsamea* (L.) P. Mill.), spruce (*Picea* spp.), white pine (*Pinus strobus* L.), and red pine (*Pinus resinosa* Ait.) from October through early December for the seasonal greens market.

Allan harvests on both private and public land. In the latter case, he indicates that regulations require boughs to be cut flush with the bole of the tree, a practice he believed was responsible for the decreasing quantity of balsam boughs available. When harvesting on land where this limitation is not imposed, Allan prefers to cut only the two most recent year's growth. This practice, he says, "causes side branches to grow more" so that he can recut from the same trees every 3 years. He indicated that branches should not be harvested from more than one-third of a tree's total height and says that he makes it a rule never to leave litter in the woods.

Bough harvesting is part of a mix of largely self-employment activities that support Allan's family of five. Because it is so critical to their seasonal income sources, it is not surprising that he was able to provide a detailed accounting of the costs associated with it. Speaking in 1996, he indicated that he spends $20-40 per day on gasoline, $21 per roll of twine (each can tie approximately 5 tons of boughs into bundles), and $14 for clippers, of which he buys 10-14 per season. In addition, he reports purchasing $200-300 worth of permits each year. At 38 years of age, Allan states that he is an especially fast worker in the woods, cutting 4 hours per day whenever the weather permits. With the 1995 buyer's purchase price of $140-160 per ton, he cleared approximately $9 per hour for his efforts.

[1]Names have been changed to protect gatherer's confidentiality

EXAMPLE 2. Knowledge–spatial and temporal distributions
Stewardship practices–minimize impacts, efforts to promote growth

Janet and Dave[1]

For more than 25 years, Janet and Dave have run a small wild edibles business out of their homestead, supplying a variety of mushrooms, leeks (*Allium tricoccum* Ait.), fiddleheads (*Matteucia struthiopteris* (L.) Todaro & spp.), and "anything edible" that grows in the UP to upscale restaurants around the country. European Americans who moved to the UP from lower Michigan when their children were young, they report gathering more than 20 NTFPs in the region.

Although not their sole source of income, the wild edibles business is their primary livelihood activity. Janet said that one objective of their early business plan was to show that it is possible to make a living in the forest without harming it. She stated that "When we get out of an area, you'll never know we were there." Janet and Dave prefer to do most of their own gathering with assistance from immediate family members and close friends. However, the need for additional labor at busy times has prompted them to train local high school students and others to harvest selected products. Janet indicates that they visit the locations where these new gatherers have harvested and will not buy from them a second time if they have not observed norms for sustainable harvest and minimal disturbance of the area.

After more than two decades of searching through UP forests for wild edibles, Janet and Dave have amassed considerable knowledge about the plants they gather. At the time of our interview, Dave had over 20 years of notes on the locations, ecological characteristics, and weather patterns of places and times that he has found edibles. He notes that leeks are one of the first green things to appear in the spring woods. In his experience, they grow primarily in hardwood stands, especially around elms (*Ulmus* spp.), and seem to like moist ground. However, he has found them both on hills and in swales and says they often are not in places he expects them to occur. "For conservation purposes," Janet and Dave have tried to propagate leeks. They tried to cultivate them on their own from seed and provided corms to a local greenhouse. Both efforts were unsuccessful, as were attempts to interest a regional university in taking on the task.

[1]Names have been changed to protect gatherer confidentiality

EXAMPLE 3. Knowledge–functional uses and preparations, harvesting for desired qualities and continued availability
Stewardship practices–time gathering, promote growth, protect sites

Walter

Walter reported gathering seven NTFPs. Among these is wiikenh[1] or flag root (*Acorus calamus* L.). He dries the root and prepares it as a tea to treat sore throats, canker sores, toothaches, headaches, stomach ailments, and gynecological problems. He indicates that spring-harvested wiikenh is weaker medicinally and less bitter; when harvested in the fall, it is more potent and bitter.

Walter is active in efforts to reconstruct and restore Native American cultural practices in Michigan's UP and his efforts in this regard have included building a birch bark longhouse. He also makes birch bark containers, which he jokingly referred to as "the original Tupperware."[2] Walter described his technique for harvesting birch bark at length. When the sap is running, a shallow horizontal cut is made around the tree at what will be the top of the piece. A second horizontal cut is made at the desired bottom, then a vertical incision is run between the two previous cuts. Walter, like other gatherers who described this process, said that when done properly and at the right time the bark pops off, often with an explosive sound. He contends that trees may be reharvested every three years with the bark becoming smoother with each harvest and with age.[3] He also stressed that it is critical to harvest only when the sap is running hard and the inner bark should never be scored.

Wild rice (*Zizania* spp.) provides another example of Walter's knowledge and stewardship practices. He harvests rice only for his household and extended family. However, because it commands a significant market price and is in high demand, he stressed that he reveals the location of his 'rice river' only to trusted family members. Walter is making an effort to increase the population of rice along rivers in his area by broadcasting seed. He harvests by boat in a two-person effort, with one poling against the current while the other knocks the seed into the craft. Walter stated that there are two harvests, an early crop in August and a late crop in September. Once gathered, he lays out the rice to air dry for a day, turning it a couple of times, then scorches it in a wash tub. A skilled tradesman, Walter thrashes and winnows his rice using a method he devised to force air through an old drum-shaped grill. He and his family consume the rice steamed, boiled, and popped like corn.

[1]The Anishinabe name is provided first as this is the name that Walter used.
[2]The use of trade, firm, or corporation names in this publication is for the information and convenience of the reader. Such use does not constitute an official endorsement or approval by the U.S. Department of Agriculture or the Forest Service of any product or service to the exclusion of others that may be suitable.
[3]Another individual reported observing a 10-year rotation length.

EXAMPLE 4. Knowledge–temporal distributions
Stewardship practices–importance of observing all norms

Abby

Abby, whose ancestry includes French Canadian and Eastern European Jewish forebears, as well as Ojibwa, Potawatomi, and Odawa, identifies herself as Native American. She listed 38 NTFPs that she gathers and has strong feelings about what she regards as the right way to harvest them. In addition to the norms listed in Table 2, Abby was the first person to mention four Native American gathering rules:

- Pick everything the way it is supposed to be picked (i.e., to assure it's continued growth and reproduction).
- Always put down tobacco as an offering.
- Ask the plant's permission before harvest.
- Tell the plant how it will be used.

These rules are regarded as particularly important when gathering plants for medicinal or ceremonial uses. Tradition also prohibits the sale of such products. However, she noted that not all Indians observe these norms. "Some gathering approaches are in conflict," she said, indicating strong disapproval when telling me about beds of the sacred plant, sweet grass (*Hierochloe odorata* L.), which she had harvested for years before they were over harvested by youth trying to make money.

Abby also described a broad temporal schematic for NTFPs. While some things can be harvested all year (e.g., spruce gum, white pine needles, spruce tips), in general "the woods dictate what you can get when." She indicated that leaf products tend to be gathered in the spring and summer, roots in the fall, and bark and whole tree items such as sap in the late winter and early spring. She and her family often scout for NTFPs and harvest them in combination with other activities such as hunting and fishing. Thus, she noted that chokecherries (*Prunus viginiana* L.) coincide with bear season and high bush cranberries (*Viburnum* spp.) with bird season, the latter being eaten together at Thanksgiving dinner.

CONCLUSIONS

The quotation with which this paper opens continues with a list of topics for which hard scientific data are lacking:

> *the economics of NTFP management, trade and marketing in different forest types; (on) biological production function for*

> *most NTFP species; local harvesting and utilization patterns; and the impacts of commercialization and changing patterns of use on the state of NTFP and related activities.* (Tewari and Campbell 1995, pp. 57-58)

In fact, Janet and Dave, Walter, Abby, Allan, and other experienced gatherers possess considerable knowledge of these facets of the products they gather. Taken individually, their knowledges are rich in specific detail of the temporal and spatial patterns of NTFPs and their livelihood uses. Examined collectively, they provide extensive information about the biological and social ecologies of NTFPs. Such local knowledges also provide valuable information on stewardship practices for sustainable NTFP use. Gatherers' harvesting norms are intended to assure the continued biological presence and harvesting availability of NTFPs. Of course, not all gatherers observe appropriate norms at all times; to paraphrase Flora speaking of farmers, gatherers are no more moral than the rest of us (1992; p. 96). But long-term gatherers have every incentive to preserve both the presence of products in the forest and their access to them. In many cases, they have had years to observe the combined effects of their harvesting and changes in weather or land use and adjust their practices accordingly.

Thoughtful combination of local knowledges and formal science has been suggested as a means of arriving at more socially and ecologically sustainable agricultural and natural resource practices (DeWalt 1994; Flora 1992; Kimmerer 2000; Murdoch and Clark 1994). There is every reason to believe that such an articulation would be valuable for sustainable NTFP use in the United States as well. For example, gatherers' stewardship practices might serves as the basis for studies of the impacts of specific harvesting techniques on NTFP populations and ecological functions. Incorporating such 'real world' information increases the potential for research to produce results that are both socially and ecologically sound.

Just how a melding of local knowledges and formal science is to be achieved is a worthy agenda for NTFP researchers over the next several years. While the specifics are likely to be as diverse as the situations in which they are implemented, researchers and managers who wish to incorporate local knowledges into their work should observe some guidelines:

- Seek experienced gatherers from a variety of age, gender, income, and ethnic groups to develop a comprehensive picture of local NTFP knowledges and practices.
- Be prepared to invest time and effort to learn from gatherers. Understand that their knowledge is rich in detail and may sometimes draw on informal experimentation and/or readings in formal literature, but is generally not expressed in systematic terms.
- Respect gatherers' right to choose what information they share and their need to protect continued access to NTFPs as livelihood resources.

Through day-to-day experience and the intergenerational transfer of skills, NTFP gatherers have accumulated significant knowledge of the products they harvest and practices that make them a sustained source of livelihood resources. Learning to recognize and work with information produced outside formal scientific channels will immediately increase the knowledge base available for NTFP research and management.

NOTES

1. No products were included for which only second-hand reports were available.
2. 1995 average annual unemployment in the region was 8.9% (Salow 1996) versus 5.6% for the nation as a whole (Bureau of Labor Statistics 1997).
3. For herbaceous products, it is particularly important to know associations with tree species because the latter are large and much easier to see from a distance.
4. Intervals varied by product and individual.

REFERENCES

Anderson, M.K. 1999. The fire, pruning, and coppice management of temperate ecosystems for basketry material by California Indian tribes. Human Ecology 27(1):79-114.

Bureau of Labor Statistics. 9/12/97. U.S. Unemployment rates, 1986-1995. In: Civilian Labor Force Survey, edited by user-defined enquiry system. http://stats.bls.gov/cgi-bin/surveymost: generated by Marla Emery.

Cleland, C.E. 1992. Rites of conquest: The history of Michigan's Native Americans. University of Michigan Press, Ann Arbor.

DeWalt, B.R. 1994. Using indigenous knowledge to improve agricultural and natural resource management. Human Organization 53 (2):123-131.

Dunbar, W.F. 1965. Michigan: A History of the Wolverine State. William B. Eerdmans Publishing Company, Grand Rapids, MI.

Emery, M.R. 1998a. Invisible livelihoods: Non-timber forest products in Michigan's Upper Peninsula. Doctoral Dissertation, Department of Geography, Rutgers University, New Brunswick, NJ.

Emery, M.R. 1999. Social values of specialty forest products to rural communities. In: Josiah, S.J. (ed.). Proceeding of the North American conference on enterprise development through agroforestry: Farming the agroforest for specialty products, Center for Integrated natural Resources and Agriculture Management (CINRAM), Minneapolis, MN.

Feldman, S. and R. Welsh. 1995. Feminist knowledge claims, local knowledge, and gender divisions of agricultural labor: Constructing a successor science. Rural Sociology 60 (1):23-43.

Flora, C.B. 1992. Reconstructing agriculture: The case for local knowledge. Rural Sociology 57 (1):92-97.

Gottesfeld, L.M.J. 1994. Conservation, territory, and traditional beliefs: An analysis of Gitksan and Wet'suwet'en subsistence, Northwest British Columbia, Canada. Human Ecology 22 (4):443-466.

Grenand, P. and F. Grenand. 1996. Living in abundance: The forest of the Wayampi (Amerindians from French Guiana). In: M. Ruiz Perez and J.E.M. Arnold (eds.). Current issues in non-timber forest products research. Center for International Forestry Research, Jakarta, Indonesia.

Haraway, D.J. 1989. Primate visions: Gender, race, and nature in the world of modern sciences. Routledge, London.

Harding, S. 1986. The science question in feminism. Cornell University Press, Ithaca, NY.

Homma, A.K.O. 1996. Modernization and technological dualism in the extractive economy in Amazonia. In: M. Ruiz Perez and J.E.M. Arnold (eds.). Current issues in non-timber forest products research. Jakarta, Indonesia: Center for International Forestry Research.

Kainer, K.A., and M.L. Duryea. 1992. Tapping women's knowledge: Plant resource use in extractive reserves, Acre, Brazil. Economic Botany 46 (4):408-425.

Kimmerer, R.W. 2000. Native knowledge for native ecosystems. Journal of Forestry. 98(8): 4-9.

Kloppenburg, J., Jr. 1991. Social theory and the de/reconstruction of agricultural science: Local knowledge for an alternative agriculture. Rural Sociology 56 (4):519-548.

Love, T., E. Jones, and L. Liegel. 1998. Valuing the temperate rainforest: Wild mushrooming on the Olympic Peninsula Biosphere Reserve. Ambio Special Report No. 9 (September 1998):16-25.

Martin, J.B. 1986. Call It North Country: The Story of Upper Michigan. Third Edition (original printing 1944). Wayne State University Press, Detroit.

McLain, R.J., and E.T. Jones. 1997. Challenging 'community' definitions in sustainable natural resource management. International Institute for Environment and Development (IIED), London.

Merton, R.K. 1973. The sociology of science: Theoretical and empirical investigations. Harper and Row, New York.

Mosse, D. 1994. Authority, gender and knowledge: Theoretical reflections on the practice of participatory rural appraisal. Development and Change 25: 497-526.

Mulkay, M. 1979. Science and the sociology of knowledge. George Allen & Unwin, London.

Murdoch, J., and J. Clark. 1994. Sustainable knowledge. Geoforum 25 (2):115-132.

Nabhan, G.P., D. House, H. Suzan A., W. Hodgson, L. Hernandez S., and G. Malda. 1991. Conservation and use of rare plants by traditional cultures of the U.S./Mexico borderlands. In: M.L. Oldfield and J.B. Alcorn (eds.). Biodiversity: Culture, conservation, and ecodevelopment, Westview Press, Boulder, CO.

Richards, R.T. 1997. What the natives know: Wild mushrooms and forest health. Journal of Forestry 95 (9): 5-10.

Salow, K.J. 1996. Upper Peninsula unemployment rates: Spring 1996 calculated benchmarks. Michigan Employment Security Commission: Information and Reports Section, Marquette, MI.

Schmidt, T.L., Jr. J.S. Spencer, and R. Bertsch. 1997. Michigan's forests 1993: An analysis. Vol. Resource Bulletin NC-179. North Central Forest Experiment Station, St. Paul, MN.

Tewari, D.D., and J.Y. Campbell. 1995. Developing and sustaining non-timber forest products: Some policy issues and concerns with special reference to India. Journal of Sustainable Forestry 3 (1):53-79.

Turner, N.J. 1998. Plant technology of First Peoples in British Columbia. UBC Press, Vancouver.

U.S. Census Bureau. 1990. 1990 Census of population and housing. http://venus.census.gov/cdrom/lookup: U.S. Census Bureau Internet Site.

Recent Trends: Non-Timber Forest Product Pickers in the Pacific Northwest

Richard Hansis
Eric T. Jones
Rebecca J. McLain

SUMMARY. The Pacific Northwest is a region where commercial demand for a variety of NTFPs–floral greens, mushrooms, berries, mosses–has expanded rapidly over the past fifteen years, creating space for new types of harvesters. These are mainly recent Southeast Asian and Latino immigrants who find this work allows them some degree of self-direction and income. Tensions have arisen between Native Americans, Euro-Americans, and recent immigrants over access rights to NTFPs as competition for these previously abundant resources has increased. Increased harvesting has also brought concerns about sustainable harvesting forward.

Richard Hansis is Coordinator, Environmental Science Program, Humboldt State University, Arcata, CA 95521 USA.

Eric T. Jones is Co-Founder and Director of the Institute for Culture and Ecology, Portland, OR 97212 USA.

Rebecca J. McLain is Co-Founder and Director of the Institute for Culture and Ecology, P.O. Box 6688, Portland, OR 97228 (E-mail: mclain@ifcae.org).

[Haworth co-indexing entry note]: "Recent Trends: Non-Timber Forest Product Pickers in the Pacific Northwest." Hansis, Richard, Eric T. Jones, and Rebecca J. McLain. Co-published simultaneously in *Journal of Sustainable Forestry* (Food Products Press, an imprint of The Haworth Press, Inc.) Vol. 13, No. 3/4, 2001, pp. 141-146; and: *Non-Timber Forest Products: Medicinal Herbs, Fungi, Edible Fruits and Nuts, and Other Natural Products from the Forest* (ed: Marla R. Emery, and Rebecca J. McLain) Food Products Press, an imprint of The Haworth Press, Inc., 2001, pp. 141-146. Single or multiple copies of this article are available for a fee from The Haworth Document Delivery Service [1-800-342-9678, 9:00 a.m. - 5:00 p.m. (EST). E-mail address: getinfo@haworthpressinc.com].

KEYWORDS. Immigrants, participation, conflict

NEW PICKERS IN THE WOODS[1]

One of the trends in recent years in the United States is the large increase in the participation of non-Euro-American ethnic groups in forest work. In the Pacific Northwest, the only area in the United States for which documentation is readily available, NTFP activities provide a large share of the forest work opportunities available to non-Euro-Americans. For most of the 20th century, most commercial pickers were predominately Euro-Americans, and rural or small town residents. Many recreational pickers, predominately Euro-American as well, dwelled in large cities. Native Americans also gathered NTFPs for spiritual and subsistence purposes and small amounts for sale. Over the past decade these demographics have shifted greatly with the entry of large numbers of immigrants from Southeast Asia and Latin America, especially Mexico and Guatemala, into forest occupations in the Pacific Northwest region.

Southeast Asian Harvesters

The first wave of immigrant harvesters consisted mainly of peoples who came as refugees from Cambodia and Laos. These immigrants began by picking wild forest mushrooms, primarily matsutakes (*Tricholoma magnivelare* (Peck) Redhead, also known as pine or tanoak mushrooms), morels (*Morchella* spp.), and chanterelles (*Cantharelleus cibarius* Fr), as well as bear grass (*Xerophyllum tenax* [Pursh] Nutt), for the floral greens industry. To a lesser extent, a few Southeast Asian groups at the time were also picking salal (*Gaultheria shallon* Pursh), ferns, tree boughs, mosses, and huckleberries (*Vaccinium* spp.). Many of these immigrants live in large and medium sized cities as well as some smaller towns on the west coast of the United States, but often come from smaller cities and villages in their country of birth. Some harvest only during certain seasons, specializing for example in the harvest of certain species of wild mushrooms. Others engage in NTFP harvesting on a year-round or nearly year-round basis, harvesting a variety of NTFPs over the course of the year.

Why have these immigrants selected NTFP harvesting as an occupation? For those immigrants with few skills applicable to the United States' economy and who do not read, write, or speak English well, NTFP harvesting provides an opportunity to earn money in a self-directed activity rather than working as a menial laborer in the city. For others, NTFP picking allows them to supplement their other sources of income. For both groups, going to the forest to pick may also serve to maintain extended family bonds because NTFP work can often be done as a family or extended family activity. For those families who enjoy forest settings or who wish to escape urban life for a short time, NTFP harvesting also makes it possible for them to afford to spend time in remote rural settings.

Southeast Asian immigrants began harvesting wild mushrooms in western Washington and Oregon in the mid-1980s. By the late 1980s and early 1990s, Southeast Asian pickers became the predominant harvesters of wild matsutake mushrooms in certain areas, notably in the eastern Cascades of Oregon and Washington and the Siskiyou Range in southwest Oregon and Northern California. They also make up an important subset of wild morel pickers in eastern Oregon. Southeast Asians also pick chanterelles in western Oregon and Washington, but they constitute a much smaller percentage of the work force in those areas.

Latino Harvesters

Latinos first entered the forest work force in the United States as reforestation workers and as workers on Christmas tree farms. Over time Latino immigrants expanded into other forest work, including the harvesting of floral greens–huckleberry greens, ferns, salal and boughs. In the past five years, Latinos have also begun to take on a more prominent role in harvesting wild mushrooms.

Some Latinos engage in NTFP harvesting as an alternative to farm work, a common occupation for new immigrants from Mexico and Central America. More recently, settled Latino farm workers have turned to NTFP harvesting for additional or alternative sources of income during poor agricultural seasons or during times of the year when agricultural work is not available owing to poor harvests or increased competition for the farm jobs. In the Cascade range of Washington and Oregon, for example, many Latino farm workers started by picking huckleberries, but now are beginning to harvest many other NTFPs.

CHANGING LABOR RELATIONS IN PACIFIC NORTHWEST NTFP INDUSTRIES

The larger numbers of potential pickers have brought with them both increased cooperation among ethnic groups and conflict between groups. For example, in the Gifford Pinchot National Forest (located in western Washington), bear grass used to be harvested in small amounts by Native Americans, primarily for use in basket weaving. In the late 1980s, bear grass became a large volume commodity, due in large part to the presence of a new labor supply consisting primarily of Cambodian immigrants. Over the years, Cambodian immigrants working in the Gifford Pinchot National Forest have begun to assume the role of labor brokers-crew foremen rather than harvesters. In this new role, the Cambodian immigrants recruit young Latino immigrants to do the heavy, sometimes cold and wet work of harvesting bear grass. Research on the nature of labor relations within NTFP industries is in the beginning stages, making it difficult to assess the relationships between NTFP employers, foremen, and crew members. However Latinos who participated in preliminary investigations of a NTFP study in the eastern Cascades indicated that their wages are very low, whether in bear grass, floral greens, or bough harvesting.

CONFLICTS OVER RESOURCE TENURE

Many Native Americans and Euro-American residents who have harvested NTFPs express resentment of immigrant harvesters, who they feel are invading their forests. Native Americans have strong legal claims to NTFPs across much of the Northwest. Their claims are strongest on lands reserved to American Indian nations under treaties signed by the United States government in the mid-1800s. Laws regarding the harvest of NTFPs on reservation lands are made by tribal governments and in many cases only tribal members and their relatives have access to NTFPs on reservation lands unless permission is given to others. Although some non-Native Americans deliberately cross reservation boundaries in search of NTFPs, in many cases boundaries between land ownerships are poorly marked and pickers may inadvertently trespass.

Native American claims to NTFPs off-reservation are also strong in many areas, since many treaties included provisos in which the signatory tribes retained rights to traditional gathering grounds. In practice this has meant that Native Americans have been able to successfully maintain or, more recently, reassert claims to NTFPs on lands managed by the federal government (e.g., national forests, national parks, and lands managed by the Bureau of Land Management). It is as yet unclear whether Native American gathering rights retained under these treaties also extend to state and private lands.

Traditional claims of Euro-American pickers to NTFPs, especially on public lands, are based on customary use rather than grounded in formal laws or treaties. For much of the 20th century, large public and private landowners either didn't regulate most NTFP harvesting activities, or didn't enforce the regulations on paper. Over the years, NTFP pickers developed informal claims to their customary gathering grounds within the context of this de facto open access property regime. Many new pickers do not know about or have not respected those claims, creating tension with older pickers.

The tensions over NTFP access are heightened by differences in harvesting practices among the new pickers, especially the Southeast Asian immigrants, and long-time Euro-American and Native American pickers. For example, the practice of raking wild mushroom beds, a technique practiced in some parts of Southeast Asia, is considered destructive by both Euro-American and Native American wild mushroom pickers. Differences in how the various ethnic groups behave in forest settings further exacerbate the tensions. Southeast Asian mushroom pickers, for example, tend to pick in large groups whereas Euro-Americans tend to pick in groups of two or three. Many Euro-American pickers feel threatened by the presence of large groups of strangers in the woods; they also feel that the presence of too many pickers in an area damages the ability of mushroom beds to fruit.

CULTURAL DIVERSITY AND LATENT CLAIMS: TWO CHALLENGES OF NTFP MANAGEMENT

NTFP harvesters in the Pacific Northwest are as diverse in terms of their cultural backgrounds and values as the products they harvest are biologically diverse. This diversity is reflected both in the ways that different groups organize themselves to participate in NTFP harvesting and in the views that they have about what constitute sustainable harvesting practices. Most media accounts of NTFP harvesters in the Pacific Northwest portray diversity in the NTFP arena as one of the major barriers to ecologically sound management. However, recent research in the Amazon rainforest suggests that cultural diversity, and the resultant wide range of accepted management practices, may be a critical element in creating and maintaining biologically diverse forest ecosystems (Hyndman 1994). Learning how to encourage and support a wide range of harvesting systems and practices, rather than developing a fairly narrow range of acceptable practices, thus constitutes one of the major challenges for sustainable NTFP management in the Pacific Northwest. The existence of multiple layers of latent tenure claims, and the conflicts that will arise as different groups seek to enforce those claims, constitute a second major challenge that will need to be addressed as the value of NTFPs, economically, culturally and ecologically, becomes more widely recognized.

NOTE

1. The discussion in this section draws from published work by Hansis (1996, 1998), Love and Jones (1997), and Richards and Creasey (1996); it also draws from the authors' collective experiences as researchers, buyers, and harvesters in the wild mushroom and floral greens sectors in Washington and Oregon.

REFERENCES

Hansis, R. 1996. The harvesting of special forest products by Latinos and Southeast Asians in the Pacific Northwest: preliminary observations. Society and Natural Resources 9:611-615.

Hansis, R. 1998. A political ecology of picking: non-timber forest products in the Pacific Northwest. Human Ecology 26(1):49-68.

Hyndman, D. 1994. Conservation through self-determination: promoting the interdependence of cultural and biological diversity. Human Organization 53(3):296-302.

Love, T. and E.T. Jones. 1997. Grounds for argument: local understandings, science, and global processes in special forest product harvesting. In: Vance, N. and T. Love (eds.), Special forest products: biodiversity meets the marketplace. Proceedings of fall 1995 seminar series. Oregon State University. Corvallis.

Richards, R.T. and M. Creasey. 1996. Ethnic diversity, resource values, and ecosystem management: matsutake mushroom harvesting in the Klamath bioregion. Society and Natural Resources 9:359-374.

Expanding Non-Timber Forest Product Harvester/Buyer Participation in Pacific Northwest Forest Policy

Rebecca J. McLain
Eric T. Jones

SUMMARY. During the past decade, a variety of new state and federal laws and regulations have been developed to regulate the use and management of NTFPs on federal and state lands. A growing body of literature on the social aspects of NTFPs indicates that few NTFP harvesters and buyers are involved in the development of these rules. This policy overview draws upon the authors' five years of ethnographic research on the politics of NTFPs and wild mushrooms in the Pacific Northwest region of the United States to describe and analyze barriers to NTFP harvester and buyer participation in NTFP policy fora. Three case examples of efforts by participants in NTFP industries to organize themselves politically so that they can have a voice in policy and management decisions are discussed. The overview concludes with a series of recommendations for steps that non-governmental organizations and public land management agencies can take to support harvester/buyer efforts to expand their influence over forest policy and management decisions.

Rebecca J. McLain is Co-Founder and Director of the Institute for Culture and Ecology, P.O. Box 6688, Portland, OR 97228 (E-mail: mclain@ifcae.org).

Eric T. Jones is Co-Founder and Director of the Institute for Culture and Ecology, Portland, OR 97212 (E-mail: etjones@ifcae.org).

[Haworth co-indexing entry note]: "Expanding Non-Timber Forest Product Harvester/Buyer Participation in Pacific Northwest Forest Policy." McLain, Rebecca J., and Eric T. Jones. Co-published simultaneously in *Journal of Sustainable Forestry* (Food Products Press, an imprint of The Haworth Press, Inc.) Vol. 13, No. 3/4, 2001, pp. 147-161; and: *Non-Timber Forest Products: Medicinal Herbs, Fungi, Edible Fruits and Nuts, and Other Natural Products from the Forest* (ed: Marla R. Emery, and Rebecca J. McLain) Food Products Press, an imprint of The Haworth Press, Inc., 2001, pp. 147-161. Single or multiple copies of this article are available for a fee from The Haworth Document Delivery Service [1-800-342-9678, 9:00 a.m. - 5:00 p.m. (EST). E-mail address: getinfo@haworthpressinc.com].

KEYWORDS. Participation, public involvement, wild mushrooms, forest policy, non-timber forest products

THE PUSH FOR NTFP POLICY REFORM

As competition for NTFPs increases in the United States, forest stakeholders including public and private forest land managers, scientists, and NTFP harvesters and buyers, are calling for regulatory and policy reforms in the NTFP sector (Schnepf 1994). Federal, state, and private land managers have become more vigilant in enforcing existing laws and regulations related to NTFP use and management. They are also developing new laws and regulations to address the conflicts that have arisen as new types of pickers enter the woods and as larger numbers of people seek to harvest larger quantities of NTFPS. However, many NTFP harvesters and buyers often have had only limited involvement in the development of these new laws and regulations. As a result, new rules governing NTFP use and management may not take into account the knowledge that NTFP harvesters and buyers have acquired over years, and sometimes generations, of harvesting these products. In addition, compliance rates are often low, as many harvesters and buyers remain unconvinced that the new regulations are based on anything more than political expediency or unreasonable fears about the dangers of overharvest.

In this policy overview, we first describe the context of NTFP policy change in the Pacific Northwest region of the United States. We then examine three examples of efforts to expand NTFP harvester and buyer involvement in the development and implementation of federal forest policies affecting NTFP use and management. We end the article with recommendations for steps that non-governmental organizations and public land management agencies can take to support harvester/buyer efforts to expand their influence over forest policy and management decisions.

Data used to develop this policy overview were gathered primarily through both authors' observations of a variety of NTFP policy making arenas in Oregon and Washington during the years 1993-1998, including McLain's participant observation as a wild mushroom buyer on a National Forest in central Oregon in spring 1997 and spring 1998, and through Jones' ethnographic study of wild mushroom harvesters in fall 1995 and fall 1996 in western Washington (Love et al. 1998).

Policy making arenas the authors observed and participated in included Washington State legislative hearings on the renewal of Washington State's Wild Mushroom Act of 1989 and modification of Washington's special forest products legislation, meetings of the Western Oregon Special Forest Products Council an ad-hoc committee composed of Bureau of Land Management and U.S. Forest Service Employees in Oregon and Washington, and annual meetings and conferences of the Northwest Special Forest Products Association, a regional association of NTFP buyers. The authors also relied upon organization newsletters, meeting notes, and key informant interviews to describe the activities and to assess the strengths, and weaknesses of the three organizations explored in the case examples.

Political ecology analyses, such as this one, operate from the assumption that solutions to natural resource problems require an understanding of the political underpinnings that shape them. In the case of Pacific Northwest NTFP industries it is clear that NTFPs have long been of significant economic importance to many families and communities, but the potential economic impacts of active NTFP management have not been seriously examined by forest managers or scientists. This paper is a first step toward understanding how existing configurations of power in decision-making have kept NTFP commercial interests from being instrumental in shaping forest policy, and in pointing out strategies that NTFP harvesters and buyers are using to try to shift the balance of power in their favor.

A SHIFT IN PUBLIC FOREST MANAGEMENT PARADIGMS

The push for NTFP policy reform on federally owned forests in the Pacific Northwest coincides with a shift in the management paradigm that shapes management priorities of the U.S. Forest Service and the Bureau of Land Management, the two largest public land management agencies in the Pacific Northwest. In the past five years both agencies have adopted ecosystem management as a guiding principle, the goal of which is to manage for the long-term integrity of whole ecosystems, not for the production of single resources (Lertzman et al. 1997). Though there is little experience to date in the U.S. with how to translate the ideals of ecosystem management into praxis, this change in official orientation requires agencies to allocate more resources to understanding and appropriately managing the extraction of NTFP resources from public lands.

CRITIQUES OF EXISTING PUBLIC PARTICIPATION OPPORTUNITIES

Federal land management agencies are also under pressure to improve opportunities for public participation in land use decisions. Federal land management agencies are required by law to solicit public comment on proposed management alternatives. However, these agencies have been criticized for structuring participation in ways that discourage significant participation by politically and economically weak stakeholders (Blahna and Yonts-Shepard 1989; Cortner and Shannon 1993; Culhane 1990). These agencies also rely heavily upon scientific knowledge to guide and justify policies and management actions, a practice that sociologists of science argue disfavors stakeholders who lack the means to participate in the production and use of scientific knowledge (Waller 1995; Rouse 1987; Love and Jones 1997). Some policy analysts and social scientists also argue that other forms of knowledge, such as experiential knowledge gained over years of working in a particular area or with a particular product or sets of products, are potentially essential contributors to sustainable resource management decisions (Allen and Gould 1986; Messerschmidt and Hammett 1998; Waller 1995).

FORGOTTEN VOICES

Recent studies of present-day NTFP harvesters in the Upper Midwest (Emery 1998) and the Pacific Northwest (Love et al. 1998; Richards 1997) illustrate the extensive knowledge that some harvesters have of both the products they harvest and the ecosystems in which those products are located. Many commercial and subsistence harvesters have been found to regularly experiment in their daily work to identify stewardship practices that would ensure the long-term productivity of the resources from which they derive their livelihoods (Love et al. 1998; McLain and Jones 1997). Yet few NTFP harvesters or buyers perceive themselves as having a voice in making new rules regarding NTFP allocation mechanisms, fee structures, and size and quantity restrictions (Kantor 1994; Robinson 1994; Love et al. 1998). On-going research of wild mushroom policy by the authors suggests that NTFP harvester/buyers also have little voice in defining and selecting more general forest management options, such as size and locations of timber sales, timing and locations of controlled burns, or location of road closures (McLain and Jones 1998). Yet these types of decisions also can greatly affect both NTFP productivity and people's ability to access key NTFP gathering grounds.

The limited involvement of NTFP harvesters and buyers in the decisions that affect their livelihoods has at least two negative consequences for sustainable forest management. First, management decisions made without significant involvement by harvester/buyers do not adequately incorporate the broad range of experiential knowledge that is available about NTFPs, and the resulting policies often mesh poorly with social and ecological conditions. Second, the regulations developed by public forest managers without input by harvester/buyers lack social legitimacy among NTFP harvester/buyers, and noncompliance rates are often extremely high.

BARRIERS TO NTFP HARVESTER AND BUYER PARTICIPATION IN POLICY MAKING

The following factors constitute some of the main barriers to harvester participation in policy making in the Pacific Northwest (McLain and Jones 1998).

- Communication is poor between different groups of harvesters and between public land managers and harvesters. Where managers have developed linkages it generally has been limited to only a few of the most visible NTFP buyers and largest NTFP businesses.
- The absence of organized groups that can legitimately claim to "speak" for NTFP harvesters and buyers as a whole, or even for specific subgroups, in informal or formal policy fora.
- Land managers often convey a disdainful attitude toward NTFP industries, harvesters, and buyers, and frequently do not accord them the same level of respect given to more numerous or more powerful stakeholders (i.e., timber companies, environmental organizations, recreationists).

- Harvesters, and many buyers and small businesses have grown distrustful of researchers who they feel extract information and tie up their harvest areas for studies without providing access to alternative sites or producing information that has a positive effect on NTFP policy from their perspective.
- Public land managers lack the proper skills to identify and reach out to marginal stakeholder groups, and to create appropriate fora for resolving conflict, exchanging knowledge, and forging mutually agreed upon solutions.

EXPANDING NTFP HARVESTER AND BUYER PARTICIPATION IN POLICY MAKING

A variety of NTFP stakeholders, including harvesters, buying and processing companies, Native American tribes, amateur scientists, professional scientists, and public land management agency staff, are engaged in efforts to overcome many of the barriers listed above. These efforts often involve the use of multiple strategies, such as facilitating the formation of political and economic alliances within and between harvester/buyer groups, supporting the development of partnerships between harvester/buyers and other stakeholder groups, and encouraging harvester/buyer participation in scientific research and inventory and monitoring. Three examples of how harvester/buyers are attempting to acquire a greater voice in policy making for Pacific Northwest forests are described below.

Trinity Bioregion NTFP Partnership, Northern California, (Everett 1997)

Beginning in the late 1980s, a group of NTFP harvester/buyers and small farmers in the Trinity Bioregion of Northern California began working together to expand their access to NTFP and organic produce markets. These initial efforts, with funding support from Trinity County and the USDA-Forest Service, have gradually developed into a partnership among two wildcrafting cooperatives (Trinity Alps Botanicals and the High Mountain Herb Cooperative), U.S. Forest Service managers from two national forests, the U.S. Forest Service Pacific Southwest Research Station, the Watershed Research and Training Center (a local non-governmental organization), and members of the Hupa tribe. Since 1995, the various partners have participated in meetings and workshops to discuss concerns related to NTFP use and management in the Trinity Bioregion. These meetings have helped create and strengthen communication links among NTFP businesses in the Trinity region and between the local NTFP industry and public land management agencies.

Practical outcomes of these workshops include the expanded marketing opportunities for independent wildcrafters and coordination in filling orders between cooperatives; the development of training workshops for NTFP harvesters and the publication of a jointly produced pamphlet which provides guidelines for sustainable harvesting of certain NTFP species. Some of the partners are also cooperating in the development of a field inventory methods and geographic information systems (GIS) based inventories of expected NTFP distributions. The latter are based upon existing data already contained in the Forest Service's GIS, which the agency uses in making its public land management decisions. Other partners have joined forces to carry out NTFP regeneration trials on public land. Silvicultural and agroforestry trials are also planned. Both types of trials will help the various partners refine sustainable harvesting and management guidelines for products harvested on public and private holdings. The partners are currently working with the Forest Service to develop a permit that will include information that harvesters can use to increase the price of their products (e.g., official notice that the local forests have been herbicide-free since 1984), and that will request information feedback from wildcrafters that can be used for ecological monitoring.

Forest Workers/Harvester Network Program, Pacific Northwest, (Jefferson Center 1996, 1997, 1998a, and 1998b)

Many commercial NTFP harvesters in the Northwest belong to economically and politically weak groups, and communication links among the different groups are quite limited. NTFP harvesters also are often only weakly connected to other forest workers, such as tree planters, stream restorationists, and vegetation survey technicians, who might share some of the same concerns over forest management and working conditions. As a result neither NTFP harvesters, nor the broader collection of forest workers, have sufficient cohesion as a group to make themselves felt in public forest policy fora.

To facilitate the formation of group solidarity among the disparate types of forest workers and NTFP harvesters, the Jefferson Center, a small non-governmental organization based in Oregon, has sponsored a series of forest worker/harvester gatherings in Oregon, Washington, and California over the past 4 years. The gatherings that have taken place thus far have brought ethnically and occupationally diverse peoples who work in the region's forests together to share and talk about their experiences in the woods. They also have served as a mechanism for the participants to begin identifying concerns that transcend occupational and ethnic backgrounds and to discuss possible options for addressing those concerns through collective action. Over time, the gatherings have evolved to include Cambodians, Mien, Latinos, European-Americans, and Native Americans. The meetings have been conducted in English, Spanish, and Khmer with the help of simultaneous translators.

In recent meetings, participants have started to identify ways in which they can take concrete action as a group to improve their ability to influence forest management decisions. One outcome was the development of a group letter addressed to the chief of the U.S. Forest Service stating that contract forest workers and NTFP harvesters need to be involved with the proposed monitoring and training programs for the agency's new collaborative stewardship program. In a meeting held in June 1997, participants voted to initiate a planning process that has led to the formation of a three-state association. In 1998, the newly-created Alliance of Forest Workers and Harvesters sponsored a series of meetings between wild mushroom harvesters and the U.S. Forest Service that resulted in the lengthening of the wild mushroom season on four National Forests in central Oregon and in the institution of a system of wild mushroom industrial camp meetings as a means to expand harvester input into Forest Service wild mushroom regulations and policies (Jefferson Center 1998a and 1998b). The Jefferson Center forest worker/harvester network program also informally links participants in the Trinity Bioregion partnership with forest workers/harvesters in Oregon and Washington. It also links participating forest workers/harvesters with the Lead Partnership Group in California and Northwest Sustainability Working Group in Washington, tying them into a wider network of stakeholders interested in supporting sustainable forest management.

Northwest Special Forest Products Association (NSFPA), Pacific Northwest (Carlson 1994; McLain 1995; NSFPA 1994a and b; Perpetual Forest Resource Newsletter, issues Winter 1994 through Winter/Spring 1996)

The Northwest Special Forest Products Association began as an informal network between several floral greens companies based in Oregon. These companies started meeting in 1992 to figure out how they could work together to improve landowner views of the NTFP industry and to influence the content of legislation and regulations developed at state and federal levels. In 1993, the group became a formal non-profit organization with members drawn from a variety of NTFP industries. In January 1994, the group received a $30,000 grant from the Oregon Economic Development Department to assist it in developing the NTFP industry, which had been designated as one of Oregon's key industries in rural areas. Full members (with voting privileges) were required to be licensed NTFP companies, while anyone interested in NTFPs could enroll as a non-voting associate member.

The association has a number of goals: to improve the image that public and private landowners have of the NTFP industry, to lobby for legislation that protects the interests of NTFP industries, and to serve as a communication link within the NTFP industry and between NTFP companies and federal agencies. During its first two years, the association sponsored several workshops that brought together NTFP businesses, researchers, and land managers. The association also helped fund the printing of Perpetual Forest Resources, a regional newsletter about NTFP harvesting and policy issues.

The NSFPA's efforts to influence policy more directly, however, have been unsuccessful. Thus far members have been unable to reach consensus on the establishment of harvesting guidelines and standards or on whether the association should support legislation requiring buyers to record permit numbers and identifying information from harvesters that sell to them. The association has also been unable to agree on whether harvesters should be permitted in as full voting members. Some buyers believe that without harvester support the association will not be able to accomplish its goals, others are worried that harvesters would quickly outnumber licensed companies and take over the association's leadership. The association's members also cannot agree on whether they should take on the responsibility of carrying out a periodic update of the price surveys conducted by university researchers in 1989 and 1996-97.

DISCUSSION

The three groups described above differ considerably in their structures and ways of operating, but are similar in that they all originated out of a desire on the part of NTFP harvesters or buyers to improve their capacity to influence policies that affect their lives and to shift the economic terms of trade in their favor. Although the particulars vary, the strategies used by the three groups to accomplish these objectives are similar:

- Each group has emphasized the importance of improving communication within the NTFP industry, particularly across subsectors;
- All three groups have made an effort to reach out to other stakeholders who share some of their concerns; and
- Outside support in the form of funds or advice has played a critical role in helping these groups get started.

How well are these groups doing at accomplishing their goals? Two of these groups, the Trinity Bioregion Partnership and the Jefferson Center Forest Worker/Harvester Network, appear to be thriving. Both have taken some concrete steps to influence policy decisions that will affect NTFP harvesting or buying activities. In addition, the Trinity Bioregion Partnership has enabled harvesters and buyers to take an active part in the scientific experiments and inventory and monitoring processes that are likely to result in guidelines that constrain their behavior. The Partnership is also working to ensure that harvesters' experiential knowledge is integrated into management and policy decisions. The Forest Worker/Harvester Network has also brought together workers from a 3-state area. This step is likely to be a necessary precursor for the development of policies that adequately address the concerns of mobile NTFP harvesters, whose gathering ranges may encompass portions of several states.

In contrast, the NSFPA appears to be on the verge of extinction. The association had some early successes, notably its co-sponsorship of a well-attended regional conference on the business and science of

NTFPs in 1994 (Schnepf 1994). However, it has not been able to maintain sufficient internal cohesion to take any collective action in the policy realm. Factors that may have contributed to the NSFPA's inability to take concrete action in the policy sphere and to its imminent demise include reliance on leaders and coordinators with little training in group facilitation skills, irregular and infrequent meetings, an unwillingness to open full membership to critical allies (i.e., NTFP harvesters), and the lack of committed and long-term support from its members and outside organizations.

ENCOURAGING AN ENABLING ENVIRONMENT FOR HARVESTER/BUYER PARTICIPATION

Although the three cases above demonstrate that harvester/buyers are already taking steps to provide themselves a greater voice in a variety of policy fora across the Pacific Northwest, evidence suggests that support from outside the NTFP harvesting and buying communities may be a critical factor in the success of these initiatives. In the next paragraphs, we briefly outline some of the steps that non-governmental organizations and public land management agencies can take to encourage an environment that enables harvesters and buyers to work more effectively in informal and formal political arenas. The type of support provided can vary, ranging from the provision of funds to sponsoring training in facilitation and leadership skills to serving as allies in legislative lobbying campaigns. However, it should be noted that outside support can also be a double-edge sword. Individuals and organizations that claim to be neutral catalysts for cultural empowerment can also use a group to accomplish hidden agendas, or more innocuously, to filter actions through the values of the organizers.

Steps That Outsider Groups Can Take to Minimize Negative Effects on Harvester/Buyers

- Understand the cultural dimensions and history of the NTFP environment before taking action.
- Consider the implications of closed meetings and selective stakeholder processes versus open public meetings.
- Build reflexive critique of the organization and activities into all decision and management processes so that the purpose and goals of the organization's actions are routinely scrutinized.

- Form relationships with third parties who can offer detailed analysis and critique aimed at improving the organization's decision-making and management processes in ways that will not violate the privacy of the members or compromise the political power of the organization.

Steps That Public Land Management Agencies Can Take to Support Harvester/Buyers

- Public land management agencies need to have staff who are knowledgeable about non-timber forest products. Many field offices do not have full-time staff members assigned to non-timber forest products management. In those that do, the NTFP program managers often know little about them or have competing responsibilities. Under these circumstances it is hardly surprising that the public agencies lack interest in NTFP harvesters and buyers.
- Public land managers can display a more cooperative spirit with harvesters by conveying positive attitudes and activities. Some concrete steps are listed below (McLain and Jones 1997).
  - Provide harvesters and field buyers with basic infrastructure such as reasonable camping facilities with showers, portable water, garbage disposal areas, and public telephones.
  - Expand hours of business for issuing permits or provide self-permitting facilities after agency business hours.
  - Post regulations and announce meetings in the appropriate languages and in places where harvesters will find them.
  - Create handbooks or other materials that explain existing avenues for harvesters to affect forest policy (e.g., how to get on mailing lists for meetings, explanation of legal rights with respect to access to resources and participation in decision making, and sources for additional aid and information).
  - Promote the participation of harvesters in research projects, designing training programs if necessary.
- Land managers can develop relationships with applied social scientists with skills in broadening stakeholder participation. The United States Environmental Protection Agency's recent cooperative agreement with the Society for Applied Anthropology provides one possible model that other public agencies can use for

accomplishing this task (EPA 1996). Through the cooperative agreement, the EPA has access to independent professionals with the methodological skills needed to profile the social dynamics, values and concerns of communities and to better understand local knowledge and priorities. These professionals also are skilled in helping communities express their conceptions of nature, risk, and change and can help facilitate interactions among contending groups so that collective action becomes possible.

Taking these steps requires more time and energy than the approaches typically used in NTFP management today. However, they have much to offer in terms of improving the quality of management decisions, increasing the rates of regulatory compliance, and helping public land agencies meet the challenges of carrying out sustainable forest management.

REFERENCES

Allen, G.M. and E.M. Gould. 1986. Complexity, wickedness, and public forests. Journal of Forestry 84(4):20-23.

Blahna, D.J. and S. Yonts-Shepard. 1989. Public involvement in resource planning: toward bridging the gap between policy and implementation. Society and Natural Resources 2:209-227.

Carlson, M.S. 1994. Meet the new Special Forest Products Association. In: pp. 163-166. C. Schnepf (ed.). Dancing with an elephant. Proceedings of the business and science of special forest products: a conference and exposition. January 26-27, 1994. Sponsored by Northwest Special Forest Products Association and Western Forestry and Conservation Association, Hillsboro, Oregon.

Cortner, H.J. and M.A. Shannon. 1993. Embedding public participation in its political context. Journal of Forestry 91(7):14-16.

Culhane, P.J. 1990. NEPA's effect on agency decision making. Environmental Law 20:681-745.

Emery, M. 1998. Invisible livelihoods: non-timber forest products in Michigan's Upper Peninsula. Ph.D. Dissertation. Rutgers University, New Brunswick, NJ.

Environmental Protection Agency (EPA). 1996. Cooperative agreement between the Society for Applied Anthropology (SFAA) and the Office of Sustainable Ecosystems and Communities (OSEC). United States Environmental Protection Agency. October 1996. http//www.telepath.com/sfaa/coop.htm.

Everett, Y. 1997. Building capacity for a sustainable non-timber forest products industry in the Trinity Bioregion: Lessons drawn from international models. Rural Development Forestry Network Paper 20a. Winter 1996/97:24-40. Overseas Development Institute.

Jefferson Center. 1996. Third annual forest workers/harvesters gathering. November 9-10, 1996. West Salem, Oregon. Summary of meeting notes.

Jefferson Center. April 10, 1997. Letter written to Chief Mike Dombeck, United States Forest Service on behalf of the Forest Harvester Network. Jefferson Center Newsletter. 1998a. Mushroom harvesters and US Forest Service discuss matsutake program. Jefferson Center Newsletter 1(3):1 and 3.

Jefferson Center Newsletter. 1998b. 1998 matsutake season a bust, but communications among harvesters, USFS, improve. Jefferson Center Newsletter 1(4):3.

Kantor, S. 1994. Local knowledge and policy development: special forest products in coastal Washington. Masters Thesis. University of Washington, Seattle.

Lertzman, K., T. Spies and F. Swanson. 1997. From ecosystem dynamics to ecosystem management. In: pp. 361-382. Schoonmaker, P.K., B. von Hagen, and E.C. Wolf (eds.). The rainforests of home: profile of a North American bioregion. Island Press, Washington, DC.

Love, T. and E.T. Jones. 1997. Grounds for argument: local understandings, science, and global processes in special forest products harvesting. In: pp. 70-87. N. Vance and J. Thomas (eds.). Special forest products: biodiversity meets the marketplace. Proceedings of fall 1995 seminar series, Oregon State University. U.S. Department of Agriculture, Forest Service, Washington DC.

Love, T., E.T. Jones and L. Liegel. 1998. Valuing the temperate rainforest: wild mushrooming on the Olympic Peninsula Biosphere Reserve. Ambio Special Report 9:16-25.

McLain, R. 1995. Participant observation notes from Northwest Special Forest Products Association 1995 annual meeting. June 11, 1995. Brownsville, Oregon.

McLain, R. and E.T. Jones. 1997. Challenging community definitions in sustainable natural resource management: the case of wild mushroom harvesting in the USA. Sustainable Agriculture Program. Gatekeeper Series No. 68. International Institute for Environment and Development, London, England.

McLain, R. and E.T. Jones. 1998. Participatory non-wood forest products management: experiences from the Pacific Northwest, USA. In: pp. 189-196. Lund, H.G., B. Pajiri, and M. Korhonen (eds.). Sustainable development of non-wood goods and benefits from boreal and cold temperate forests. Proceedings of the international workshop. Joensuu, Finland: 18-22 January 1998. European Forest Institute Proceedings No. 23, 1998. European Forest Institute, Joensuu, Finland.

Northwest Special Forest Products Association (NSFPA). August 21, 1994a. Bylaws of the Northwest Special Forest Products Association, Inc.

Northwest Special Forest Products Association (NSFPA). November 1994b. Special Newsletter.

Perpetual Forest Resource Newsletter. Winter 1994, Spring 1994, Fall 1994, Winter 1995, Spring 1995, Summer 1995, Fall 1995, Winter/Spring 1996.

Richards, R.T. 1997. What the natives know: wild mushrooms and forest health. Journal of Forestry 95(9):5-10.

Robinson, C. 1994. Multiple perspectives: rules governing special forest product management in coastal Washington. Masters Thesis. University of Washington, Seattle.

Rouse, J. 1987. Knowledge and power: toward a political philosophy of science. Cornell University Press, Ithaca, NY.

Schnepf, C. (editor). 1994. Dancing with an elephant. Proceedings of the business and science of special forest products, a conference and exposition. January 26-27, 1994: Hillsboro, Oregon. Sponsored by Northwest Special Forest Products Association and Western Forestry and Conservation Association.

Waller, T. 1995. Knowledge, power, and environmental policy: expertise, the lay public, and water management in the western United States. The Environmental Professional 17: 153-166.

Synthesis and Future Directions for Non-Timber Forest Product Research in the United States

Rebecca J. McLain
Susan J. Alexander

SUMMARY. During the past decade, NTFPs have begun to appear on mainstream scientific research agendas in a variety of disciplines. Development of a strong NTFP research capacity will require the construction of links between on-going and emerging NTFP research programs focused on U.S. NTFP issues, establishment of strong ties to international NTFP research programs, and the use of interdisciplinary and collaborative research approaches. Understanding forests as biophysical systems that also include humans will enhance the effectiveness and relevance of U.S.-oriented NTFP research efforts. *[Article copies available for a fee from The Haworth Document Delivery Service: 1-800-342-9678. E-mail address: <getinfo@haworthpressinc.com> Website: <http://www.HaworthPress.com>]*

KEYWORDS. NTFP research, science

NTFP research in the United States is in its infancy and participants in such projects are struggling to develop approaches that will prove

---

Rebecca J. McLain is Co-Founder and Director of the Institute for Culture and Ecology, P.O. Box 6688, Portland, OR 97228 (E-mail: mclain@ifcae.org).

Susan J. Alexander is Research Economist with the USDA Forest Service Pacific Northwest Forest Sciences Lab, Corvallis, OR 97331 (E-mail: salexander@fs.fed.us).

[Haworth co-indexing entry note]: "Synthesis and Future Directions for Non-Timber Forest Product Research in the United States." McLain, Rebecca J., and Susan J. Alexander. Co-published simultaneously in *Journal of Sustainable Forestry* (Food Products Press, an imprint of The Haworth Press, Inc.) Vol. 13, No. 3/4, 2001, pp. 163-165; and: *Non-Timber Forest Products: Medicinal Herbs, Fungi, Edible Fruits and Nuts, and Other Natural Products from the Forest* (ed: Marla R. Emery, and Rebecca J. McLain) Food Products Press, an imprint of The Haworth Press, Inc., 2001, pp. 163-165. Single or multiple copies of this article are available for a fee from The Haworth Document Delivery Service [1-800-342-9678, 9:00 a.m. - 5:00 p.m. (EST). E-mail address: getinfo@haworthpressinc.com].

successful in the long run. Programs with the capacity to address the complex socio-ecological questions that are being raised about NTFP use and management in the United States and elsewhere will need to have the following characteristics:

- *Stronger Links Between NTFP Research Programs in the United States:* In trying to put together a team to write this issue, it became clear that while links between NTFP researchers in the Pacific Northwest are relatively strong, few of the scientists working in that region have close connections with scientists working on NTFP issues in other parts of the United States. Yet market conditions, and thus pressure for resource use, for at least some products are tightly connected to supplies and demands for those products in other parts of the country. Additionally, lessons learned about how to support widespread involvement of the various types of NTFP stakeholders in policy and management in the Northwest might have applicability to other parts of the country. Lack of regional comparisons also makes it difficult to identify those aspects of NTFP use in the Pacific Northwest that are geographically and historically specific. This creates a risk of inappropriate generalization to other areas of the United States, with potentially unfortunate implications for policy development by national land management agencies. Collaboration and information exchange among research programs across the United States could help researchers make better use of scarce funding and human resources.
- *Stronger Links to International Research Programs:* An annotated bibliography on NTFPs in the Pacific Northwest region of the United States (von Hagen et al. 1996) indicates that the issues associated with NTFP use and management in that region are similar to NTFP issues in many other parts of the world. Although the specifics may differ from place to place, some of the basic questions are similar: (1) How does one set up inventory and monitoring programs for species with high temporal and spatial variability? (2) What types of accounting methods can be used to adequately assess the economic value of NTFPs in forest ecosystems? (3) How does one incorporate both scientific and experiential knowledge about NTFPs into forest management decisions? (4) What can be done to ensure that a wide range of

NTFP stakeholder interests are adequately represented in political processes? Collaboration and information exchange between NTFP programs in the United States and NTFP programs in other countries could prove mutually beneficial for addressing these questions.

- *Interdisciplinary, Collaborative, Multi-Stakeholder Approaches:* Many of the questions related to NTFPs cut across disciplines and social institutions and are important to a wide variety of stake holders. Additionally, funding for NTFP research has been scarce and large increases in funds and human resources, particularly within governmental research and land management institutions, are unlikely to be available in the near future. Programs that pull together scientists from a variety of disciplines, that leverage scarce funds and human resources through collaborative agreements between various social institutions, and that encourage participation by a wide variety of stakeholders are thus likely to be important elements in the creation of a long-term, wide-spread NTFP research capacity in the United States.

Non-timber forest products have been important to people who harvest them for personal use or for sale for a long time. Forest managers are sometimes surprised at the demand for NTFPs. For example, over 11,000 personal use permits for berries and mushrooms were issued in 1996 on the Gifford Pinchot National Forest in Washington State. As the contributions to this issue illustrate, research on NTFPs in the United States is addressing a wide variety of topics including ecosystem management, biology and yields, market and recreational demand and value, tribal rights, and harvester behavior, knowledge, and demographics. Seeing forests as whole systems that include human beings as well as other biological and physical components will take us a long way toward avoiding the mistakes of the past as we seek to understand and manage non-timber forest products.

Index

CPSIA information can be obtained
at www.ICGtesting.com
Printed in the USA
BVHW040437271119
564913BV00007B/49/P

9 781560 220893